A NEW LITTLE ICE AGE HAS STARTED

How to survive and prosper during the next 50 difficult years.

Cover

The cover of this book shows the alignment of Jupiter, Saturn, Uranus, Neptune and the Earth, with the Sun on February 24, 1803, just two years after Sir William Herschel, Britain's Astronomer Royal published a paper discussing the nature of sunspots, their effect on our climate and the possibility that planetary alignments might have something to do with them. The Dalton Sunspot Minimum of the Little Ice Age started around 1800, and lasted until 1850. An alignment very similar to 1803 occurred in 2000. Sunspot counts have fallen from then until now and snowfall and cold weather are both increasing.

Cover design by Hindsight Productions.

1

This book is dedicated to all the people and in particular, the homeless, who will suffer because of the failures of their governments, the U.N., Legacy news media and the green corporations.

Published by Lawrence E. Pierce, Box 82, Hornby Island BC Canada V0R 1Z0

Copyright 2015 by Lawrence E. Pierce © First published May 22, 2015.

All rights reserved. No part of this book may be reproduced or transmitted in any form or by any means, electronic or mechanical, including photocopying, recording, or by any information storage and retrieval system, without permission in writing from the author, except for brief quotations in critical articles or reviews.

ISBN 10 1515158519 ISBN 13 9781515158516

Other books by Lawrence Pierce:

The Art of Fixing Things, principles of machines and how to repair them. Available only on Amazon.com

A big thank you to Margit Lieder for her advice, and red pen work.

CONTENTS

Introduction	7
Solar system cycles that bring cold and warm weather to Earth.	12
What happened during the last little ice age.	22
How this new little ice age might unfold.	28
Saving yourself and others.	38
Who concealed this coming disaster and what must be done about it.	46
Conclusion.	67
Appendix : Back to the land	70
End notes.	107
Index.	125
About the author.	132

Introduction

I got interested in the whole topic of 'global warming' and 'climate change' late in life. After 23 years successfully practicing litigation law in Vancouver, B. C. I took my life savings in 2003 and bought a farm in the Gulf Islands of B.C. I wanted to grow grapes, and I knew that the location was a bit cool for what I wanted to do, but, no worries: there was a scientific consensus that it would get warmer. All the newspapers said so, and back then I still believed in the newspapers.

After two fairly good crops, it got cold and the grapes grew like crazy and needed lots of pruning, but did not ripen. I began to wonder what was going on. By this time I had built a winery, and had gotten into debt.

After doing a little research, I noticed several authors talking about a new cold spell that was just around the corner. Because I had to decide what to do about the farm: sell and move south; graft different grapes onto the root stock; go into cabbages and goats, I decided to go all out to try and figure out what was happening. My inquiries were built on what I had already read, but when the winter of 2013-14 came along I got worried.

This book is the result of that research.

Many people have come to regard global warming as something they 'believe' in. The topic has taken on the status of a quasi-religion, but I prefer facts and evidence. This is my attempt to deal with the topic of global climate on the basis of science, not belief.

All the evidence available today proves that a New Little Ice Age is upon us, and it is going to be much colder and snowier in most parts of the Northern Hemisphere for the next 50 to 80 years than the winter of 2014-2015. But don't panic, knowledge is power, there are ways to survive and prosper during tough times, and it will be warmer than before in some parts of North America. Both you and the society you live in can come out of this stronger. The fact that you are reading this book is a great start.

The winter of 2012-2013 broke records all over the world. The Great Lakes were 92% frozen over, and winter seemed to drag on forever in Canada.

The winter of 2014-2015 was predicted by many researchers to be especially cold and snowy. Even the Old Farmer's Almanac predicted in the late summer of 2014 the coming "Refriger-nation" cold spell.

There is a lot more to a New Little Ice Age than ice. In the last one, the Dalton Minimum, the worst of which lasted from about 1800 to 1850, there were devastating floods, excessive rain, major wind storms, rapid temperature changes, heavy snow and bitter cold in North America, and Europe, and a serious increase in volcanic activity around the world. However, there were also nice days, and generally the weather in the Western United States, and the West Coast of Canada *was warmer than parts of North America* in the 1800s and warmer than the 20th century.

Sunspot cycles have been recorded for hundreds of years. Best known is the eleven year cycle. There is a complicated way to count them, but once counted, they can be put into a chart, and the eleven years runs from peak to peak. There are slight variations in this cycle, like in any natural cycle, because this cycle is based on the periodic alignment of the planets.[2]

Longer cycles than the eleven year cycle are much more important. How do sunspots affect us? Actually they have no effect on us at all, but serve as a warning of what is happening inside the sun. Low counts have been associated with cold weather on Earth since their discovery by Galileo in 1609, with a new invention: the telescope.

In the Nineteenth Century the price of grain on the London market was high when sunspots were low, and when there were lots of sunspots the price of grain was low. This suggests that low sunspot numbers predict cool weather, but it takes more than a simple correlation, without more evidence to be so bold as to predict today that a catastrophic climate change to cold weather is upon us. These sorts of correlations are useless for predicting anything.

More recent scientific enquiry has demonstrated that when sunspot numbers are low, there is decreased output of solar magnetism, or the 'solar wind' which normally protects the Earth from harmful cosmic rays. The Sun's heliosphere surrounds and protects the entire solar system from these rays when the sunspot number is high.

There is a solid connection through observation of physical clues between low sunspots, and cold and erratic weather which has a record going back hundreds of years. Past increases in cosmic rays left a trace in the oceans and trees. Carbon 14 deposited in tree rings increases under cosmic ray bombardment, and deep ocean sediments and well as ice core studies from stable ice sheets in Greenland also document the increase of these rays.

Current global warming theories are driven by computer models which have consistently made predictions that did not come true. Commenting on 73 climate model predictions that are wrong, Dr. Roy Spencer said:

"In my opinion, the day of reckoning has arrived. The modelers and the IPCC (the U.N. body that promotes global warming) have willingly ignored the evidence for low climate sensitivity for many years, despite the fact that some of us have shown that simply confusing cause and effect when examining cloud and temperature variations can totally mislead you on cloud feedbacks (e.g. Spencer & Braswell, 2010). The discrepancy between models and observations is not a new issue...just one that is becoming more glaring over time."

"It will be interesting to see how all of this plays out in the coming years. I frankly don't see how the IPCC can keep claiming that the models are "not inconsistent with" the observations. Any sane person can see otherwise."[3]

The current way of thinking about climate change is that it is human caused and all the solutions therefore must be man-made, and of course very expensive. Trillions of dollars have been spent trying to reduce carbon dioxide output, and regulate every aspect of human life. Yet the concentration of CO_2 in the air has risen from 350 parts per million (ppm) to over 400 ppm in the last 60 years, with no increase in warming in the last 18 years.

The events of the last 200 years point to the conclusion that a new Little Ice Age has started. It is based on records of sunspots, backed up by other hard data including several hundred years of actual temperature measurements. This growing body of scientific evidence is much more reliable, because it is based on observations not on computer models, which are really just predictions of how the future might turn out.

The U.S. Government and the EU continue to push global warming. Fortunately, both Canada and Australia have taken a different approach. Canada withdrew from Kyoto, and is obviously committed to not harming the economy with questionable carbon dioxide cap and trade schemes. The government of Australia fell recently over carbon taxes, and was replaced by a party that is working for Australians, not against them.

The predictions of changes in the Sun, the 'ultimate source' of climate,[4] have been around a long time, some going back hundreds of years. This is true science. Make an observation, and/or come up with a theory, publish it, see if anyone else agrees, and in the case of theories related to the natural world, see if it can be verified by more observation.

Contrary to what some people and groups have been saying, our planet has been in a cooling phase for decades.[5]

Dr. Jack Sauers a Washington geologist said in 1999:

"The last 650 years have also seen a lowering of the Douglas firs' tree line by 1,000 feet in the Cascade Mountains of Washington state, as evidenced by the U.S. Forest Service's thousands of test plots. Their test plots in the Olympic Mountains have also showed a slowing of tree growth, which is evidence of a strong increase in global cooling, throughout the 1990s."[6]

Here is a chart of data compiled by Rutgers University Global Snow lab that shows an increasing trend in snow coverage in the Northern Hemisphere from 1967 to 2014.[7]

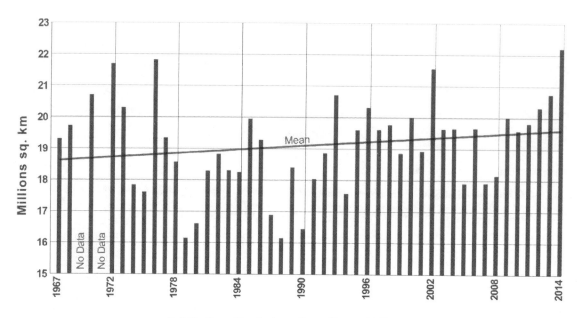

Fall Northern Hemisphere Snow Extent 1967 - 2014

The same is true in Europe. To determine the spring trends, German researchers looked at the mean temperature for February in Germany, which is a country that is ideally situated in Central Europe. Cold and snow at the end of February have a considerable impact on when vegetation starts to blossom. What follows is a chart depicting the February mean temperature for Germany over the last 28 years. The temperature is progressively colder each year.

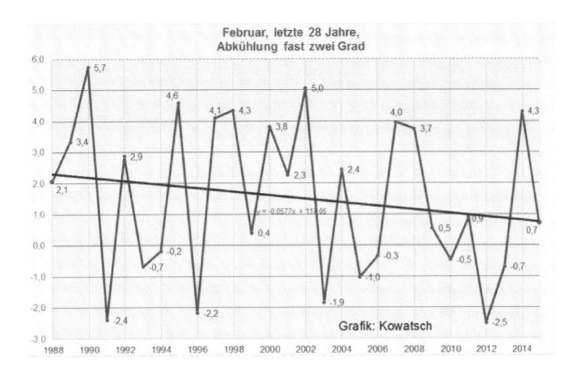

"After almost 30 years of winter cooling we see: In the open unbuilt areas of Germany, where today there are no longer any weather stations, the following remains valid: Spring awakens in March. Spring awakens in March as it did 150 years ago at the end of the Little Ice Age." [8]

1. Solar system cycles that bring cold and warm weather to Earth

The movement of planets around the sun causes changes in the sun and changes in protection for Earth. There is more than one theory about how often these cycles come around, and how long they last. As well, there are different lengths to these solar cycles, and because the orbit of the Earth around the Sun changes, the impact of cycles varies.

"Eleven year" cycles are the shortest, and range from 7 to 17 years. There is some evidence that they are caused by a conjunction or alignment of the Earth, Mars, Venus and Mercury, or some of them with the Sun.

The next recognized cycle is 200 years, although some experts say it is 178 and some say 206. Then comes the double 200, which seems to be recognized as a 360 or 400 year cycle, and finally there is the 1440 (or 1430) year cycle which is four times 360.

Ian Plimer says: "The recognition of 1430-year periodicity in the volcanic and climate records, now called the Dansgaard-Oeschger Cycle, is not new. It was first recognized in 1914." [9]

But the most important thing is that, just like a boomerang, what happened in the past will come back again, causing climate disruptions in varying strengths and for varying durations.

A solar minimum is the moment at which the number of sunspots bottoms out, and heads back up. It will be a moment that will last for a few weeks or months. A solar maximum is just the opposite, the sunspot number tops out, then falls.

A solar minimum and the months and years following are associated with very cold weather, as well as very strong winds, rain, unusual low pressure areas, and many other disturbances. There have been two recent and notable historical periods with decades-long episodes of low solar activity. The first is known as the 'Maunder Minimum', named after solar astronomer Edward Maunder, and it lasted from 1645 to 1715. The second one is referred to as the 'Dalton Minimum', named for the English meteorologist John Dalton and it lasted from 1780 to 1830. Together they are referred to as the "Little Ice Age". 1810 was the only year in a two hundred year span with no sunspots at all. [10]

There have been many solar minima, all running on approximate 200 year cycles. These are their names, and the approximate start date for each.

Oort 1000

Wolf 1250

Sporer 1400

Maunder 1645

Dalton 1780

Eddy 2000

Sunspots have been observed and counted since the telescope was invented in 1609. Galileo Galilei was the first to record these, and other celestial objects. They can be seen with a specially filtered telescope and are cool regions on the sun. From our perspective, they are mainly indicators of what is going on inside the Sun. The Sun is not entirely understood, but we do know that there is a great deal going on. Heat, visible light, infrared light, x-rays, magnetic forces and gravitational forces plus rare solar flares which can fry satellites and power grids.

The Sun, our star is 93 million miles away, and its gravitational forces act on other planets like our Moon's gravity acts on the Earth to cause tides. But the reverse is also true; when several planets get together the combined gravitational pull can cause changes in the Sun's structure and behavior. Jupiter and Saturn are the principal ones, with the other 'heavy' planets, Uranus and Neptune contributing. The more planets move close to each other (conjunction), the more pull they exert on the Sun. It is also expressed as torque, which may depend on conjunction plus the movement around the Sun and other planets. [11]

The exact mechanism of planetary pull on the Sun is still being studied, but the frequency of the reaction of the Sun, in producing low sunspot counts is well understood.

The counting and naming of these sunspot cycles can be done in different ways, with the result that sometimes the count is different depending on who is doing the counting. All but the last two cycles, (Dalton and Maunder) and the current one are reconstructions from "proxies". A proxy is just some other physical record that confirms sunspots or cosmic rays.

These proxies include the isotope Carbon 14 (C14) which is found in tree rings, deep sea sediment cores, ice core counts, and even rainfall .

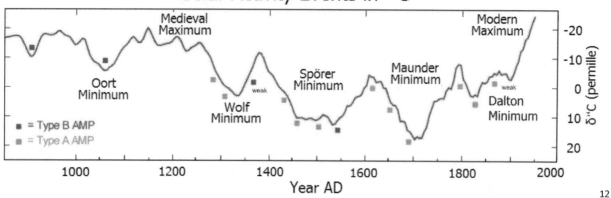

The C14 is higher in trees in years when there are reduced cosmic rays. Cosmic rays are repelled by the solar wind/solar magnetism, which is at its lowest during a solar minimum, and thus does not have the power to prevent a massive increase of cosmic rays striking the Earth. [13]

Ice and sediment cores are examined for things like the remnants of ancient sea creatures that liked cold or warm oceans, and the thickness of ice layers showing warm or cold weather in the past.

The other differences with the dates of the onset and ending of these cycles is human error, and disagreements about how to count the spots, and discrepancies between actual counts and proxy counts, and attempts to correlate the weather with a solar minimum or maximum. Further complications are that the Pacific Ocean masks the onset of cold, while the Atlantic will magnify it especially if the North Atlantic Oscillation is in a positive mode.

There is also a whopper cycle that rolls around every 1440 years which is four of the 360 cycles.

According to Professor Jack Sauers, a research geologist at Columbia University:

"The 360-year Little Ice Age cycle ...correlates with the fall of the Roman Empire. It correlates with the fall of the Sumerian Empire. It correlates with the fall of the Ottoman Empire...It correlates with the fall of the Greek Empire. And it is now coinciding with the collapse of several modern-day empires." [14]

This 1440 year cycle brings on even more dramatic and rapid changes in climate than occurs with a 200 year cycle.[15] "A similar 1440 year cycle has been found in North Atlantic deep sea sediment cores, and we are due for another one now". [16]

Ian Plimer:

"When sunspot activity collapses, the Earth cools dramatically to a Grand Minimum, a phenomenon that has occurred many times over the last 10,000 years. Because of the lag between sunspot activity and the Sun's great conveyor belt, most astronomers now predict the return of a quieter Sun. The decreased solar activity would result in increased cosmic radiation attacking the Earth, resulting in increased cloudiness. Low level clouds reflect the Sun's energy back into space, resulting in cooling of the Earth." [17]

Brian Fagan:

"The record of history shows us that...climate change is almost always abrupt, shifting rapidly within decades, even years, and entirely capricious. The Little Ice Age climate was remarkable for its rapid changes. Decades of relatively stable conditions were followed by a sudden shift to much colder weather, as in the late seventeenth century, 1740/41, or even the 1960s." [18]

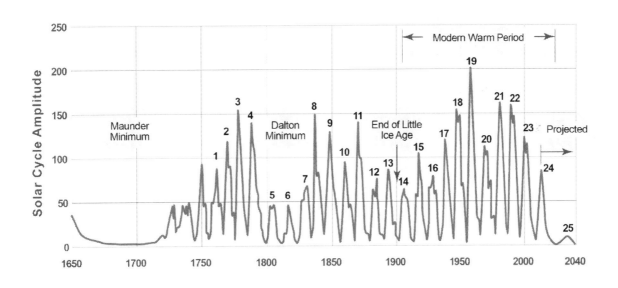

Solar Activity Cycles 1650 - 2040

We are currently just past the top of Cycle 24.

Much has been written, recently, about the cycles of cold and warm weather on Earth. There has always been plenty of data to support the idea that natural variations in climate are a normal thing, and that only the biggest egos on Earth would believe that our activities could influence the weather.

In a series of articles for the "National Post," Canadian environmentalist Lawrence Solomon interviewed several leading solar scientists, and later incorporated those interviews in a book. [19]

One of the scientists mentioned was Rhodes Fairbridge. Dr. Fairbridge reports that the subject of planetary influences on the sun, sunspots, their variability and effect on climate had been discussed as early as 1801 in a paper published by the Astronomer Royal, Sir William Herschel.

Dr Fairbridge was a legend in his field, authoring or coauthoring over 100 books and 1000 scientific papers. He was the editor of eight major encyclopedias of specialized scientific papers in the atmospheric sciences.

He concluded that sunspot activity is controlled by our solar system's center of gravity. That center of gravity is most affected by the large planets, Jupiter, Saturn, Uranus and Neptune, and when all four of these planets are on one side of the Sun, the solar system's center of gravity (the barycenter) can be altered by hundreds of thousands of miles. This shift has an effect on sunspots, and Dr. Fairbridge predicted, before his death in 2006 that the next cool period would begin in 1996, with the effects to be felt starting in 2010. [20]

More recently, Lawrence Solomon interviewed Dr. George Kukla who stated that: "We are certain now that the changes in the Earth's orbital geometry caused the ice ages. The evidence is so strong that other explanations must now be discarded or modified." [21] Dr. George Kukla is a research scientist at Lamont-Doherty Earth Observatory of Columbia University, and the author of many scientific papers.

A Russian scientist, Dr. Habibullo Abdussamatov, the head of the Space Research Laboratory at the St. Petersburg's Pulkovo Astronomical Observatory was of the opinion in 1997 that there are two solar cycles that relate to the climate on Earth, one is eleven years long, and the other is about 200 years. He also points out that the oceans store heat, so cooling is disguised by the release of heat from the oceans. [22]

Dr. Abdussamatov was also of the opinion in 1997 that the Earth had hit its temperature ceiling and is on the brink of a severe cooling that would start between 2012 and 2015, and last until 2055-2060 after which the temperature will rise again. [23]

From a 2009 scientific paper first published in Russian, Dr Abdussamatov says:

"Existence of the 11-year and 2-century solar cycles of identical and synchronized variations of luminosity, sunspot activity and diameter of the Sun is one of the most reliably ascertained facts in solar physics."

"The climate of the Earth have always been periodically changing and our planet have already experienced several global warmings, similar to the one we observe. The global warmings have always been followed by deep coolings within regular 2-century cycles. Neither a deep cooling nor a global warming cannot last longer than they are permitted by the corresponding 2-century variations of the size and luminosity of the Sun." [24]

He said: "A global freeze will come about regardless of whether or not industrialized countries put a cap on their greenhouse gas emissions." [25]

He updated his opinions in March 2015 in a paper published in the peer reviewed scientific journal "Thermal Science," coming to the same conclusion: Earth is now entering its 19th little ice age. [26]

The Russian government was so impressed with his work that he was appointed to oversee all the solar experiments conducted on the International Space Station. Quite a contrast to the attitude of the present American government who are still beating the drums of global warming. There must be more money to be made in America before more rational matters can be discussed.

In February 2014 U.S. Secretary of State John Kerry said to students in Indonesia that 'climate change' (he meant global warming) was as big a threat as terrorism, and that 'climate deniers' were supported by 'shoddy science,' and that we are somehow related to the now defunct 'Flat Earth Society.' [27]

When I was in law school, I was told that when the opposing lawyer runs out of good things to say about her client, the other lawyer would be attacked. John Kerry just demonstrated that point.

It has not been just the scientists with years of research behind them, and heavy CVs who have come to the same conclusions about the 178 to 200 year cycle. Two "non-scientists" have reached similar conclusions about the big freeze that has just started.

John L. Casey, a former space shuttle engineer, a national space policy adviser to the White House and Congress, is one. He has written: "Cold Sun" which is based on his own research. [28]

Mr. Casey says concerning the solar cycles:

"The best known is the eleven-year solar cycle, also called the Schwabe cycle. …..it varies, with some cycles as short as seven years and some as long as seventeen years. During the Schwabe cycle, the number of sunspots reaches a high point, a maximum, and then drops to a low number, a minimum, and then reaches its next high point eleven years later (on average). There are many other solar cycles and most are much longer, are far more powerful, and are *driven by the intricate movements of the Sun, the Earth, and the other bodies in our Solar System."* (Emphasis added). [29]

He goes on to summarize his own research concerning the long cycle which has the biggest effect on our climate, which he calls the Bi-Centennial Cycle, or the 206-year cycle. His conclusion in 2011 was that the next 206 year cycle change was imminent. [30]

Mr. Casey also reported on the opinions of 29 additional professional researchers and scientists working in this field, from all over the world who individually or as a team of three or four came to the same conclusion: the next change was imminent.

They made the following statements:

- "Earth has passed the peak of its warmer period and a fairly cold spell will set in quite soon, by 2012. Real cold will come when solar activity reaches its minimum, by 2041, and will last for 50-60 years or even longer."
- "In two years (after 2008) or so, there will be a small ice age that lasts for 60-80 years. "
- "This (the prediction of a very weak solar cycle) should be a great strategic concern in Canada because nobody is farming north of us."
- "It follows from their extrapolations for the 21st century that a supercenturial solar minimum will be occurring during the next few decades…It will be similar in magnitude to the Dalton Minimum, but probably longer as the last one."
- "We conclude that the present epoch is at the onset of an upcoming local minimum in long-term solar variability."
- "The surprising result of these long-range predictions is a rapid decline in solar activity, starting with cycle 24. If this trend continues, we may see the sun heading towards a 'Maunder' type of solar activity minimum…"
- "Contrary to the IPCC's speculation about man-made warming as high as 5.8 degrees C within the next hundred years, a long period of cool climate with its coldest phase around 2030 is to be expected."
- "We can say there is a probability of 94 percent of imminent global cooling and the beginning of the coming ice age."

- "These (findings) lead us to conclude that solar variability is presently entering into a long Grand Minimum, this being an episode of very low solar activity, not shorter than a century…The maximum will be late (2013.5) with a sunspot number as low as 55". [31]

David Archibald is another researcher without any university and taxpayer funds to back him up. He just got interested in cycles of cold weather and did his own research. His conclusions were later confirmed by a team of Norwegian scientists. His main thesis is that the length of one sunspot cycle predicts the length and severity of the next cold spell. [32]

Mr. Archibald's book, "Twilight of Abundance why life in the 21st century will be nasty, brutish, and short." was published in 2014. [33]

In it he makes the point that we all have a reason to thank the global warming alarmists, Al Gore, and the U.N. IPCC (Intergovernmental Panel on Climate Change.) Because if they had not been pushing so hard, many people, himself included, would not have looked at this problem, and "As a consequence, decades of discovery have been shortened into just a few years. Without the outside interest drawn in to this field of science, humanity would be sleepwalking into the very disruptive cooling (and food-production catastrophe) that will be caused by solar Cycles 24 and 25 over the years to at least 2040." [34]

Of course if the IPCC had been doing its job, we would have been alerted long before now.

Mr. Archibald's opinion, supported by the Norwegian team is that Cycle 24 has peaked, and the Sun is headed into at least two weak cycles, so that Cycles 24 and 25 will be similar to the weak Cycles 5 and 6 during the Dalton Minimum, which means extreme cold, wet and unpredictable weather across the Northern Hemisphere. [35]

Cosmic rays.

The mechanism connecting cold weather with low sunspots is as follows. The Sun's lower magnetic field strength allows more cosmic rays to bombard Earth. Cosmic rays cause cloud formation, by stimulating water vapor to form nuclei which turn into clouds, [36] and clouds reflect sunshine back into space. Clouds reflect 60 % of the Suns radiation, while open water reflects only 5%. [37]

There are other effects of changing magnetism on the Sun. These relate to the changes in the Jet Stream causing cold air to go much further South in the US, and further into Europe. [38]

Henrik Svensmark, a leader in the study of cosmoclimatology talks about cosmic rays in his book: "The Chilling Stars A Cosmic View of Climate Change":

"...physicists could see changes over thousands of years in the performance of the Sun, as the chief gatekeeper of the cosmic rays. Its magnetic field protects us by repelling many of the cosmic rays coming from the Galaxy, before they can reach the Earth's vicinity. " [39]

There is further evidence that cosmic rays have an effect on high and low pressure areas, causing, among other things, low pressure areas to stall in one spot, causing torrential downpours locally. The Calgary Flood of 2013, and the rains in Europe in 1315, during the Wolf Minimum [40] are examples of this.

In a recent study of weather and cosmic rays in the journal "Annales Geophysicae" , the author, L.I. Dorman talks about climate change, and says: "... the most important of these factors are cosmic rays and cosmic dust through their influence on clouds, and thus, on climate." [41]

Scientists at the University of Reading in the UK confirm the link between increased cosmic rays and the jet stream deformation that has occurred in the last two winters. It is the 'Polar Vortex' that will keep coming back. Increased cosmic rays are the historical link between cold winters and reduced solar output.

There is also another mechanism creating cold winters in Europe. Ozone develops in the upper atmosphere from increased cosmic rays which causes heating of the air. This hot air affects the Jet Stream. "The jet stream, which flows eastwards from North America, can readily become blocked over the Atlantic Ocean, preventing mild westerly winds from reaching Europe and enabling cold air to move down from the Arctic instead." [42]

Mr. Archibald reports on two fairly old studies which touch on the 200 year cycle, and the coming cold era.

In the late 1960's the German Navy commissioned a report on changing wind and sea conditions in the North Sea. A report issued by Weiss and Lamb in 1970 pinpointed the 200 year recurrence tendency of wind, and linked it directly to solar cycles. [43]

A second study done by Libby and Pandolfi in California in 1979 was based in part on the rings of Redwood trees. They predicted continuing cool weather through the mid 1980s, followed by severe cooling starting in about 2000. They were of the opinion that the cooling during the 21st Century could be as much as 3 to 4 degrees F, [44] which would take the world back to the coldest weather of the Dalton Minimum, during the Little Ice Age.

Mr. Archibald got curious about the changing climate, so starting with research done in Northern Ireland that suggested a link between the length of sunspot cycles and temperature, he set out to find existing temperature records that would fit with known sunspot cycles, and reviewed a number of temperature records in Europe and the North Eastern US.

Archibald says:

"For all of these locations, there is a strong correlation between solar cycle length and temperature over the following solar cycle. Projecting into the future, the European locations...have a 1.5 degree C decline in prospect on average over Solar Cycle 24. And the US locations can expect even steeper drops in temperature, with an average fall of about 2.1 degrees C." [45] Further that Solar Cycle 25 will be even longer, taking the world temperatures back to the coldest levels of the Little Ice Age. [46]

Our planet has been cooling for decades. These scientists all agree, here it is in their own words:

- "Satellite measurements reveal that the earth's lower atmosphere has been cooling for decades, says University of Alabama meteorologist John Christy"
- "Average temperatures in the United States have been falling since 1966, says meteorologist Rich Tinker of NOAA's Climate Prediction Center. It has gotten cooler in the east—especially in the lower Midwest and Southwest."
- "Tree-ring-density studies substantiate these findings. According to a 1998 study...average temperatures in the Northern Hemisphere now stand at their lowest point since 1836—and are trending strongly lower." [47]

The evidence points to the most likely explanation for both the eleven year, and (approximate) 200 year cycles being the passage of the planets around the sun. When several, including the dense ones: Jupiter, Saturn, Neptune and Uranus 'bunch up' on one side their combined gravitational pull affects the Sun, and solar magnetism, gravity and most importantly, cosmic rays striking Earth are changed. These changes affect our climate.

I also believe that the movement of the planets towards the conjunction of those four planets starts the process of solar changes, which starts the process of change here on Earth. We may not notice it at first, so when it arrives it seems sudden.

2. What happened during the last Little Ice Age?

First of all, it was not all ice. The weather varied, with some very hot summers, very cold winters, lots of wind storms caused by unusual low pressure areas, and long periods of rain and some places severe long lasting drought. One year in France might be good for agriculture, while in England it was not. Wet springs and early frosts reduced the growing season, and bitter cold often killed plants that should have overwintered. Many people died, but most survived, even though they lived in (by Western standards) primitive societies, there was no governmental social safety net, and no organized grain reserves.

Glaciers in the mountains would advance, pause, retreat, and advance again. This was truly a global phenomenon. Glaciers in the Alps would advance rapidly, then stop or retreat, crushing all in their path. The same was going on in New Zealand with the Franz Josef Glacier which: "…retreated steadily until about 1893, when a sudden forward thrust destroyed the tourist trail to the face. In 1909, advances of up to fifty meters a month were reported." This was followed by more advances and retreats. [48]

Europe in general was hit by the Black Death, which was caused in part by poor harvests, hunger and malnutrition. In the Paris region the population fell by 66% between 1328 and 1470, during the Wolf Minimum. [49]

Much bad weather in Europe can be attributed to fluctuations in the North Atlantic Oscillation which brought excessively wet springs, summers and falls, destroying much of the cereal crops that peasants relied on for food. [50] The Oscillation is either positive, or negative, and refers to climate switches between a low pressure area near Iceland and a high pressure area near the Azores. These normal oscillations were made much worse because of low solar activity.

Greenland had been settled by the Norse for 500 years before the Sporer Minimum helped to bring their culture to an end. The Greenland Norse were able to survive up to about 1408 by keeping cows, and farming the inhospitable land along the fjords .

The last written record from Norse colonies on Greenland was about a wedding on September 14, 1408. [51] Trade with Europe was also very important, but that came to an end when sea lanes became so plugged with ice that no ships could get close to the coast, or land. [52]

The end of the Norse colonies is not as simple as the weather getting cold and everyone freezing. The native Inuit survived, but for complicated reasons the conservative, Eurocentric Norse did not. Their depletion of environmental resources, failure to adopt Inuit hunting technology, as well as increased hostile contact with the Inuit [53] all combined with the Little Ice Age to force them to eat their prized cattle and hunting dogs, and survive to the end on rabbits and birds. No one from Europe visited until 1576, and when they did they found only one skeleton in a barn. The last man standing had no one to bury him.

The wealthy classes in Greenland, the Norse Chiefs were not protected by their wealth, they "found that they had merely bought themselves the privilege of being the last to starve," [54] and of course the privilege of watching everyone else starve/freeze to death. Will it be any different this time? Food will become the ultimate weapon, and the best way to defeat your enemies without looking like the bad guy is just find a way to not ship food to them.

In what would become the U.S. the colonists at Jamestown Virginia had a rough start. They arrived in 1607 at the beginning of the Maunder Minimum. The Crotan were the local inhabitants, they were an agricultural culture, and were experiencing great difficulties because of a drought which had been ongoing for 6 years. The drought ended in 1613, and everyone suffered. Of the original 104 colonists, only 38 were still alive a year later. They died of malnutrition, and water shortages as well as disease. [55]

In Canada, the Prairies had only been recently settled in 1800. European settlers tried their hand at cultivating wheat, starting during the Dalton Minimum, with very mixed success. When crops failed, they could hunt buffalo for food. The Little Ice Age was causing farming problems on this side of the Atlantic.

"The whole continent was very wild and harsh at that time. The land was overgrown, uncultivated and difficult to tame by the early European settlers. They found themselves under continuous attack by a seemingly hostile nature armed with an endless assortment of powerful natural weapons such as pests, plant diseases (rust, mould, rot), storms, floods, and rapid temperature changes."

"The earliest record of wheat cultivation in Western Canada is connected to the arrival of the Selkirk settlers in 1812."

"In a letter to Lord Selkirk dated 17 July 1813 and preserved in the National Archives in Ottawa, Miles Macdonell, the governor of the settlement, writes: "The winter wheat crop was completely wasted because it was planted too late. The same thing happened with the spring wheat, pea and English barley crops.""

"Their luck was no better the next year: the harvest of 1814 also failed. However, the persistent Scotsmen did not give up and their third attempt to grow wheat resulted in a decent harvest."

"A wheat crop of a size that would allow exports was just a dream, both for the pioneer farmers and for the government. This dream was to be fulfilled decades later with the appearance of Ukrainian wheat in Canada. It arrived at a small farm in Otonabee, Canada West, in 1842" [56]

This wheat, also known as 'Red Fife" withstood the ravages of wheat rust that destroyed other varieties Canada is a major grain supplier to the world. From 2001 to 2011 Canada exported an average of 26,330,000 tonnes of grain per year. [57]

It took the early settlers over 30 years of trial and error during the last Little Ice Age to find seed varieties that would grow consistently in this harsh climate. Many modern strains of wheat have come from Red Fife, but have been naturally modified over the years, for better yields, and higher fertilizer inputs.

Irish potato famines are well understood in North America. It is said that because of them, there are more people of Irish descent in New York City today than in Ireland.

The Irish potato famines which mainly occurred during the Dalton Minimum can be summarized as follows:

"...a terrible famine in the exceptional cold of 1740 and 1741, when both the grain and potato crops failed...livestock and seabirds (were killed)"

"More than 65,000 people in Ireland died of hunger and related diseases in 1816, the "year without a summer."

1846-1848 over 1.5 million died and 1 million emigrated because of the complete failure of Ireland's major food crop. [58]

In this same period, 1816 to 1817 thousands more died of cold, hunger and disease. The catastrophic harvest failures brought disaster on a global scale.

"Its effects ranged from the Ottoman Empire, to parts of North Africa, large areas of Switzerland and Italy, western Europe and even New England and eastern Canada. The crisis was due not only to failed harvests but also to soaring food prices at a time of continued political and social unrest after the Napoleonic wars." [59]

One lasting, and fun result of the 'year without a summer' was that Mary Shelley was visiting Lord Byron at his cold drafty castle near Geneva. It was even colder outside, and she had nothing to do so, on a bet with her husband Percy, wrote "Frankenstein." He also wrote something, but she won the bet.

The poor, as always, were hit hardest, with hunger and disease. Glasgow had 3,500 deaths and 130,000 sick people from typhus. The poor houses were overwhelmed and many had to sleep in the streets. By 1817/18 Ireland had 850,000 infected with the epidemic. [60]

One feature of the last Little Ice Age was floods. History records that really bad weather, heavy snowfall, torrential rains and winter coastal storms contributed to floods in which hundreds of thousands died.

"Storm activity increased by 85 percent in the second half of the sixteenth century, mostly during cooler winters. The incidence of severe storms rose by 400 percent. From November 11 to 22, 1570, a tremendous storm moved slowly across the North Sea…Enormous sea surges cascaded ashore, breaching dikes….By November 21 dykes were breached, drowning at least 100,000 people." [61]

"Normal or predictable spring and autumn flooding was increasingly replaced by large-area and intense flooding, sometimes outside spring and autumn from about 1300, in recurring crises which lasted into the 18th century. In the Low Countries and across Europe, but also elsewhere, the cooling trend starting in the late 13th century became more intense. It brought long cold winters, heavy storms and floods, loss of coastal farmlands, and huge summer sandstorms in coastal areas further damaging agriculture. Climate historians estimate that major flooding on an unpredictable but increasingly frequent basis started as early as 1250. Extreme events like the Grote Mandrake flood of 1362 (during the Wolf Minimum), which killed at least 100,000 people became darkly repetitive." [62]

Volcanos

Most of the writers mentioned in this book, John Casey, David Archibald, David Du Bayne, Robert W. Felix, Brian Fagan and Dr. Ian Plimer have connected earthquakes and volcanic eruptions with cold periods. There is scientific agreement that seems to relate to the solar wind, which increases during periods of reduced sunspots and reduced solar output, causing these volcanic increases.

Mt. Tambora in southeast Asia erupted in 1815, during the Dalton Minimum, spreading ash all over the world. It caused what has come to be known as "the year without a summer" in 1816. It snowed in Eastern Canada and New England in June. Crops took a beating, and more people starved the following winter.

"Estimates of climate-related deaths in North America range from the tens to hundreds of thousands, while countless more died, as epidemics of "famine-friendly" diseases such as cholera and typhus spread." [63]

If a major volcanic eruption occurs, during this current minimum, one that would put large quantities of dust into the stratosphere, it will be time to act quickly. Volcanic dust is very effective at reflecting sun light, and during the three years or so that it takes to settle out, the temperature of the earth may be lowered by one degree or more during that time. [64]

Obtaining adequate stocks of food for each family will create shortages and conflict. The 'survival of the fastest' will be the order of the day.

Another major eruption occurred during the Sporer Minimum in 1600, when the Huanyapurina volcano in Peru blew up, and caused wide spread livestock and human deaths and the destruction of the wine grape harvest for several years. The ash from this event was found in the Greenland ice cores. This eruption caused the summer of 1601 to be the coldest in the Northern Hemisphere since 1400. [65]

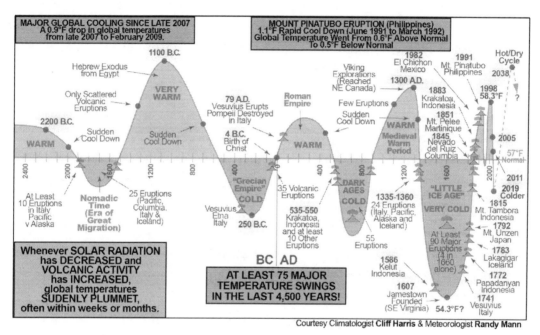

Global Temperatures 2500 B.C. - 2040 A.D.

[66]

"An almost eruption-free period from 1912 to 1963 coincided with an average global warming of 0.5 degrees C.." [67] This warm period was not very stimulating for volcanos, while today there are dozens erupting. [68]

Most volcanos emit sulphur gas, which was responsible for thousands of deaths. When the volcano Laki in Iceland erupted in 1783-84 (at the end of the Maunder Minimum) most of Europe was covered in a dry sulphuric acid fog. [69] "This eruption almost completely depopulated Iceland." [70] Millions of people died in Europe.

"There appears to be no relationship between the size of an eruption and the amount of sulphur belched into the atmosphere. A number of very minor volcanic eruptions may add more sulphur gases to the atmosphere than a massive eruption." [71]

David Du Byne has produced a number of excellent videos on the New Little Ice Age, at his "Adapt 2030" You Tube channel. His research on volcanoes and other aspects of this problem is tremendous and should be watched. He says that a reconstruction of past sunspot minima (there have been 5 in the last 1000 years) compared with major volcanic eruptions leaves no doubt about the connection.

In this video he points to the number of eruptions on going right now saying that they have already affected the weather. He predicts a major eruption in 2015. The volcanoes on land are perhaps 20 percent of the total, and when the undersea volcanoes blow up, vast amounts of water vapor is discharged into the air, providing water for rain and snow in far greater amounts than normal. [72]

A volcano on the China/North Korea border is showing signs of a possible eruption. It last blew up during the Dark Ages, and sent ash around the world, contributing to the cold weather. It is one of the biggest volcanoes in the world, on the same scale as Mt. Tambora, and could easily bring on another 'year without a summer.' [73] But do not panic yet, the experts think it might take a decade or two to blow up. [74]

3. How this New Little Ice Age might unfold.

Unusual and record breaking weather did not just begin with the "Polar Vortex" in 2014, it started several years before in both the Northern and Southern Hemispheres, and includes:

- The great blizzard of 1996, which smothered the U.S. East Coast under record snow.
- The winter of 1997-98 saw huge snowstorms in Jerusalem, Mexico City and Guadalajara; the worst ice storm in memory in Montreal, and Moscow had its coldest December since 1882.
- In 2000, 43 states in the U.S. had their coldest November and Decembers on Record.
- Mt. Baker in Washington received 100 feet of snow in one season. A new world record.
- Inner Mongolia had so much cold and snow in 1999-2000 that three million cattle died. The snowfall was 10 times heavier than normal, and in the winter of 2000-01 another 1.3 million cattle died, many of them frozen standing upright. [75]
- November 2007, Switzerland had its heaviest snowfall in fifty-two years
- August 2007, Zurich had its largest daily rainfall in a hundred years.
- March 2007, China had its heaviest snowfall in fifty-six years
- January 2007, Bangladesh had a cold wave, the coldest in Forty years.
- May 2007, South Africa reported fifty-four cold weather records, and in June 2007 Johannesburg received its first significant snowfall since 1981.[76]

The speed of the shift into the new little ice age has increased. This winter (2014-15) was brutal in North America, and the most of the rest of the Northern Hemisphere. The news media, hooked as they are on 'global warming,' did not seem to report anything more than local cold and snow…it was hard for the media to just ignore 2 to 8 feet of snow on roads and driveways, or record breaking cold. One website did report on cold and snow records that were broken around the world. [77]

In many cases, especially in March 2015, the record was not broken by one or two degrees. It was broken in some cases by 10 to 20 degrees. [78]

As of April first 2015, ferries were still having trouble getting to Newfoundland because of ice. Grocery store shelves were empty, and 350 trucks loaded with food waited at the docks in North Sidney N.S. for the ice to move. [79] As many as 15 large freighters got stuck in the ice on Lake Superior in the second week of April, and one suffered a hole below the water line from chunks of ice as big as a pickup truck, and up to eight feet thick. [80]

Here a sample of the places where cold and snow records in 2014-2015 (some going back to the beginning of records, and many going back to the mid or late 1800s.) were broken in North America:

- Cincinnati OH
- Washington DC
- New York
- Boston
- Maine
- Springfield MO
- Nashville TN
- Erie PA
- Prince Edward Island Canada
- 14 cities in Ontario Canada
- Kitamatt Village and Kitamat BC
- Terrace and Kelowna BC
- South Bend IN
- Texas Panhandle
- Amarillo TX
- Oklahoma
- Michigan
- Chicago IL
- Parts of New Mexico and Colorado
- Grand Prairie Alberta
- Pachuca City about 55 miles north of Mexico City
- Sioux Falls SD
- Buffalo NY
- Syracuse NY
- Virginia and West Virginia
- Toronto ON
- Louisiana
- Alabama
- Key West FL
- Dallas Fort Worth TX
- Jackson MS
- Little Rock AK
- International Falls MN
- Iowa
- Cuba,
- Honolulu HI.
- North Bay Ontario

States of Emergency declared in: Upstate New York, Massachusetts, St. John New Brunswick, West Virginia, and Ohio because of heavy snow fall. [81] Icebergs appeared on the beach at Cape Cod in early March. [82]

These are the record breaking locations for the rest of the Northern Hemisphere:

Jerusalem, Athens Greece, Troodos Cyprus, Istanbul Turkey, Tibet, Spain, Sweden, Norway, Sa Pa Vietnam, Saudi Arabia, Cycladic Island of Tinos Greece, Romania, Zurich, Sofia Bulgaria, Himachal Pradesh India, Pyrenees in France with possibly a world record for snow: 157 inches in three days, Bogota Columbia. [83]

March 5, 2015 a World Record for snow in one day just West of Rome, Italy at Capracotta. 8 feet fell in 18 hours. [84]

Normally typhoons occur in the summer months in the Pacific Ocean, however in 2015, each of January, February and March there was a typhoon, with significant low pressure and huge storm surges. Watch this video, which also shows snow in Mexico. [85]

Ice accumulation at the poles is up significantly, and the Great Lakes were 92% frozen in 2013-2014, and were 88% ice covered in late January 2015, 45% above median ice coverage. [86]

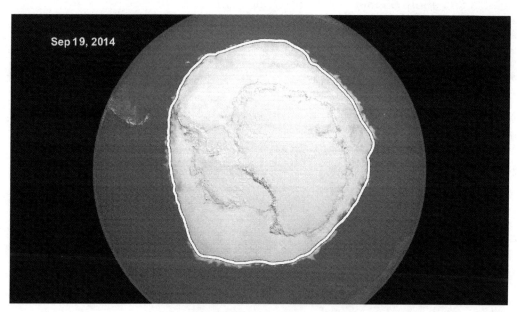

Antarctica: NASA September 19, 2014

"On Sept. 19, 2014, the five-day average of Antarctic sea ice extent exceeded 20 million square kilometers for the first time since 1979, according to the National Snow and Ice Data Center." [87]

Here is an excerpt from a NASA statement on February 27, 2015:

"The eastern half of the United States has been trapped in a deep freeze for most of February 2015. Hundreds (maybe thousands) of records have been set for daily low temperatures, and wave after wave of ice and snowstorms have hit the region. The coldest days of the year usually occur in January, yet February 2015 has been exceptional in many places." [88]

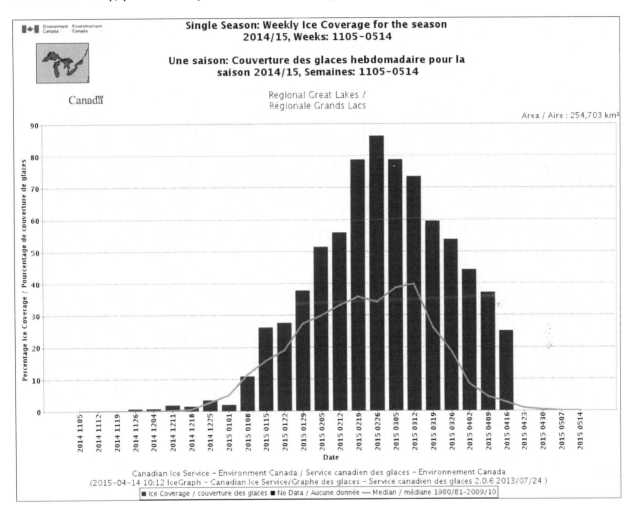

This is a chart of the ice coverage on the Great Lakes, as at April 16, 2015. The solid line is the median from 1980-81 to 2009-10.

In the last Little Ice Age weather was extreme, but:

"A modern European transported to the height of the Little Ice Age would not find the climate very different, even if winters were sometimes colder than today and summers very warm on occasion, too. There was never a monolithic deep freeze, rather a climatic seesaw that swung constantly backwards and forwards, in volatile and sometimes disastrous shifts. There were arctic winters, blazing summers, serious droughts, torrential rain years, often bountiful harvests, and long periods of mild winters and warm summers. Cycles of excessive cold and unusual rainfall could last a decade, a few years, or just a single season. The pendulum of climate change rarely paused for more than a generation." [89]

"Climate change varied not only from year to year but from place to place. The coldest decades in northern Europe did not necessarily coincide with those in, say, Russia or the American West. *For example, eastern North America had its coldest weather of the Little Ice Age in the nineteenth century, but the western United States was warmer than in the twentieth.*" (emphasis added) [90]

Warm weather records were broken or tied in the 2014-2015 winter, but mostly in the Southern and Western U.S.:

- Key West, Florida
- Miami, Florida
- West Palm Beach, Florida
- Fort Myers, Florida
- Vero Beach, Florida
- Punta Gorda, Florida
- Venice, Florida
- St. Petersburg, Florida
- (tie) Tampa, Florida
- (tie) Orlando, Florida
- Lakeland, Florida
- Plant City, Florida
- Los Angeles, California
- San Diego, California
- (tie) Brownsville, Texas
- (tie) Long Beach, California
- Phoenix, Arizona

- (tie) McAllen, Texas
- (tie) Daytona Beach, Florida
- Needles, California [91]

It was the warmest February in Vancouver B.C. in at least 60 years, if not all time. [92]

The news from Hornby Island B.C. is that we had a very mild winter and the daffodils were up by January 20, and bloomed in late-February.

During the next 50 years, we can all expect the unexpected. Some warm dry summers, early frost, reduced growing season, harsh winters, heavy rains…you name it, but none of us can name the time. The biggest risk to the world is hunger. Crops, especially grain crops are fairly sensitive. Just a little snow or rain in September on the Canadian Prairies, and the crop yield is reduced. A late spring or late frost will reduce the harvest.

Very cold weather will damage the winter wheat crop, with yield reductions from 5% to 20%.

We can also expect to see changes in our forests, over this century. Dr. Jack Sauers a Washington State geologist reports on a pollen study done south of Lake Nipigon in southern Ontario which revealed the changes over a 650 year time span. "The forest there used to be a temperate forest, beech and maples. The maples died out and gave way to oaks, then the oaks died out, and gave way to white pines. Now the white pines are disappearing and being displaced southward, and all that's coming back is boreal forest, not a temperate forest. The boreal forest is birches and aspen. Those are characteristic of what grows way up north in Scandinavia."

"So, in the last 650 years, southern Ontario has gone into the boreal plant zone. For them, the Holocene [the most recent geological warm period] is over!" [93]

As of March 2015, the FAO (Food and Agriculture Organization of the U.N.) reports world grain stocks at 600 million tonnes. With the populations of the MENA (Middle East and North Africia) countries rising every year, even small decreases in production will 'eat up' those reserves fairly quickly. [94]

World grain reserves are at about 600 million tonnes is only enough grain based on the 'stocks-to-use ratio' to feed the world for 71 days. Obviously all the crops will not fail in the same year, but a loss of say ten percent in North America, followed by another 5% the next year at the same time Europe suffers a 10% reduction will overwhelm the reserves in a very short time. Local distribution problems will compound the hunger problem. The winter of 2013-14 in Canada saw a big slow down in rail deliveries of grain to the Port of Vancouver because cold and snow required trains to run slower. The Great Lakes sea lanes were opened late in 2015, and ships had big problems with ice in 2013-2014.

The Federal Government got involved passing short term rules that would fine rail companies $100,000 per day if they did not meet delivery quotas. Bad weather can cost a lot. [95]

Rail lines these days are also plugged with tanker cars because of the failure of the U.S. Government to approve the Keystone XL Pipeline. Another global warming policy decision that will impact grain deliveries from Canada to the world, and will affect Americans in the long term.

The current projections call for a 1.5 to 2.1 degree C drop in temperature over the next 5 to 10 years, and much more in Cycle 25. This will devastate North American agriculture.

"A study of the Canadian wheat belt…found that a 1 degree C decline in temperature would reduce the frost-free period by fifteen days. A 2 degree C decline would be enough to keep the wheat crop from ripening before the first frost."

"A 1980 study of the impact of changing temperatures on the Corn Belt (in the U.S.) found that a change in temperature would shift the growing conditions by 144 kilometers(South) per one degree C. With a two degree fall coming over the next ten years, the Corn Belt will shift almost three hundred kilometers south." [96]

Hopefully North America has enough good land to change course and grow other crops, and customers who will buy those other crops. Our exports will suffer, and the countries of the world that receive our grain will suffer massive starvation. There are 500 million people in the MENA countries and almost half of their sustenance comes from imported grain. [97]

Canada is experimenting to find varieties that will grow in the New Little Ice Age, and produce large enough crops to at least keep Canadians fed. It might be possible to genetically modify wheat to withstand snow and cold, but even that will take years to produce enough seed for crops large enough to feed the world. [98]

This is especially a problem because many countries, including China, Japan, and several in Europe will not allow some types of GMO imports. The ongoing problem is producing enough seed in time to feed the world. A little advance warning might have pushed this research along faster.

One crop failure will be hard on poor countries and peoples, and two failures will spell catastrophe. Canada and the U.S. wheat and corn producers have a monumental task ahead to overcome 25 years of global warming hysteria. Additional scorn and criticism deserved or not comes from the consuming public, who see GMO foods as being a potential disaster. [99]

Further complicating the food issue is that the lead time to develop a new cold resistant strain of wheat and grow test plots can be anywhere from 10 to 12 years, with the industry trying to get it down to 6 to 7. But that is not the end of the struggle. It takes additional time to grow crops of 'foundation seed' from 'breeder seed' and more time to grow 'certified seed,' and finally large amounts will have to be grown to seed some or all of the 500 million acres around the world currently producing wheat. [100]

Presently, thousands of varieties of wheat and other cereal seeds are available for breeders to experiment with including some varieties from northern Canada and Siberia that have good cold tolerance. The limitations include getting enough certified seed that will withstand not only changes in temperature, but possible changes in rainfall across vast areas, while producing enough tonnage to feed all the grain customers of the world.

Perhaps hunger and/or high food prices will soften the criticism of GMO food.

Right now the following countries cannot even begin to feed their populations with their own produce:

- Afghanistan, with 30 million people can only grow enough to feed 13 million, in ideal circumstances.
- Egypt has 84 million and can feed 40 million.
- Yemen has 25 million and a population growth of 2.9 percent per annum but needs 9 million metric tons of grain each year. It is already in a state of civil war, and the end will come quickly if grain from the West stops flowing.
- Tunisia has 10 million people, and the population is growing at one percent per annum. They import 66 percent of their grain.
- Saudi Arabia is home to 26 million who live on grain, 90% of which is imported. Their current losses on low oil prices will probably be a guarantee of a price increase, and civil unrest when the grain and/or the money runs out
- Jordan has 6 million people who eat 96 percent imported grain.
- Syria's population is presently falling, and those left after the civil war will continue to eat 60 percent imported grain.
- Iraq's 32 million people eat 60 percent imported grain.
- Iran imports 7.5 million metric tons of grain to feed its 75 million. [101]

It does not take much imagination to realize the implications of severe grain shortages in a region grown fat on imports, where various tribes kill each other when they are not hungry, and several of these countries have or will shortly have nuclear weapons.

Nova Scotia has seen farmed fish mortality because of super chilled water. The wheat crop has already been affected, with losses of 1.5 million metric tons in Russia, and Australian exports are down 5% because of cold. Other cereal crops are also affected in Russia. In the U.S. there has been winter kill of wheat, because of low temperatures, although the extent of the losses in cereals will not be known exactly until the fall of 2015. [102]

Floods

There have been record floods in Canada, Europe and the U.K. over the last 20 years. Not all floods can be blamed on an increase in cosmic rays, caused by reduced solar wind, however floods were a regular feature of past solar minima, and today they seem to be getting more frequent and more destructive. When floods are considered with all the other evidence of a new little ice age, the argument becomes more persuasive.

Canada had four of its five worst floods since 1997, and there seems to be a common thread in the last three: excessive snow fall and rain on frozen ground.

Manitoba 2009 A heavy rainstorm in the first week of November, 2008 was a major factor in the flood. The high level of ground frost, due to the cold winter, kept the ground from absorbing much of the spring runoff.

Above-average snowpack in the US (the Red River rises near the headwaters of the Mississippi and flows North) part of the watershed also contributed to the flood.

Manitoba 2011. The 2011 flood featured the highest water levels and flows in modern history across parts of Manitoba and Saskatchewan. Southern Manitoba was within a millimetre of having its wettest year on record, when in October 2010 a super-charged weather bomb dumped 50 to 100 millimetres of rain and snow. At freeze-up, soil moisture levels were the second highest since 1948; only 2009 had more.

Cold temperatures throughout the winter resulted in deep soil-frost penetration, meaning spring melt water was likely to spread out instead of soaking in. At the season's midpoint, the snowfall total was at a 15-year high. [103]

Calgary 2013 . The largest flood in the City's history also affected one quarter of the Province and losses ran to $6 Billion. It was blamed on a slow moving storm that dumped torrential rains, as much as 200 mm (7.8 inches) of rain onto already soaked soil, and soil in some places that was still frozen. The large winter snow pack was a huge contributing factor.

"It began snowing in southern Alberta before Thanksgiving 2012 and didn't stop until a month after Easter. The mountain snowpack in May was immense, over one meter in places. Further, the spring was wet leaving the ground saturated and streams and rivers bloated." [104]

In 2013-14 record floods devastated many parts of Europe, including:

Central Europe. The worst flood in 500 years, where heavy rains were blamed for swollen rivers and flooded historic cities. In Germany people living along the Elbe river were dealing with their third flood since 2000. [105]

Germany. 2013 worst flood since the flood of 2002, which was then said to be the 'flood of the century.' [106]

The Balkans. 2014 the worst floods in 120 years.

"(In three days), three months' worth of rain has fallen on the Balkan region, producing the worst floods since rainfall measurements began 120 years ago. The flooding has affected huge parts of Serbia, and more than a quarter of Bosnia's 4 million people: The Bosnian government is warning of "terrifying" destruction comparable to the country's 1992-95 war. At least 40 people have died in the region, and tens of thousands of others have fled their homes, packing into buses, boats, and helicopters - many of them farmers leaving their livestock behind. There's an additional hazard: leftover land mines. Floodwater has disturbed known minefields and damaged some markers that had been placed there to warn people away. [107]

On March 30, 2015 Atacama Chile was the scene of a monster flood. Seven years of rain fell in 12 hours, in the driest spot on Earth. Was cosmic ray increase to blame because of the reduced solar wind, or was this just a 'freak' rainstorm? [108]

Because of the moving Jet Stream, cold air is flowing further south, creating both floods and droughts in places as far apart as Taiwan and Kashmir. The California drought may be another sign of the changing climate, because it is getting colder.

Old Spanish Colonial weather records going back to the middle of the Dalton Minimum set the benchmark for cyclone storm tracks near the Philippines and the current tracks match with the ones from the past. [109]

4. Saving yourself and others.

This book is dedicated to all the people who will suffer because we were lied to. I intend to use the sale of this book in part, for fundraising purposes to help the most disadvantaged among us: the homeless and hungry. These campaigns will be run through crowd funding sites. Watch for them, and encourage others to join in.

As Robert Felix said in "Not By Fire but By Ice"

"Tell us the truth. Put me on a plane to Alaska, tell me in advance where I'm going, and I'll figure out what to do—people will figure out what to do—we're not dummies. But tell me I'm going to Miami, then dump me off in Alaska, and I'm in trouble" [110]

First of all, there is no stopping this onset of cold, unpredictable and snowy weather. It is a reality, it is here, and the best all of us can all do is get ready. They branded people like me "Deniers" for a reason. To make our opinions less valuable, and stifle dissent so that the plan for control of everything (or whatever the plan was) could go ahead, and it almost succeeded. Now, we are all left to prepare for the worst. This is your Darwin moment.

Do not be deceived by those who are continuing to sell 'global warming' and 'climate change' deceptions. They are doing it for the money, or individuals are doing it because of the extraordinary propaganda over the last 20 years. (I am not saying this to try and sell you something...you already bought it!)

The global warming worker ants work on things like: green energy projects, 'research and mitigation to lessen the blow of a warming planet' and pushing the U.N. sponsored "Sustainability and Resilient Cities." [111]

Any person who says that man-made global warming will more than offset the effects of a little cooling is relying on all those faulty computer models that said we would be toast by now. See Endnote 2.

An organization called ICLEI is the local agent for the U.N. Their job is to help implement Agenda 21 and bring on the wealth re-distribution, and transformation of our society that the U.N. seeks. ICLEI is in every town and city. Ask your City Counselors if they provide any funds or training to the planning department. That is how they subvert property rights. Anything with the word "smart" in it must be subjected to special scrutiny. To find out what else they are really up to, one must do some research.

Agenda 21 is a treaty signed by Canada and the U.S. (and most other countries of the world) after the 1992 Rio summit, where all this was cooked up. It is under assault in the U.S. because of the planning ideas that infringe on property rights. Alabama became the first state to ban it. [112]

Some politicians, like the Governor of Massachusetts will continue to mislead us. He confidently said in February 2015 that 'next year will not be so bad'. Sorry, but it might just be worse. Politicians will try to blame every disaster on 'global warming', like Bloomberg and Cumo with Hurricane Sandy. [113]

Then there is the lazy news media...they are still talking about "global warming" saying things like: "There is a difference between climate and weather, and you can have climate change and global warming all at the same time." I wonder if they ever stop to think about how silly that sounds.

Reporters in the Lame Stream or Legacy Media, basically just re-type press releases and remind us of 'global warming.' This is a total failure of the function of the media. There is very little critical enquiry these days, the national media has chosen to report the narrative of global warming, rather than the news. The gap left by this failure has been filled by independent researchers on the internet, and the major culprits, CNN and MSNBC have been punished with low ratings. The popularity of TV is in steep decline.

The "Wall Street Journal" and the "National Post" are the only major newspapers in North America, that I know of who have consistently reported the news about 'global warming' with criticism and alternate viewpoints.

If you can ditch these old ideas, focus on your home and family. Start spending your excess cash on making your home safe and warm. Put money into lots of insulation, double pane glass, weather stripping, and consider buying a snow blower, even if you have no driveway. It can be used to blow snow up around your house to better insulate it, especially on the North side. If you have a flat roof, use a snow blower to get the snow onto the ground.

Enclose South facing porches with glass to capture as much heat as possible. Add heat saving substances, as simple as drums of water, to create a Trombe wall. Maybe build a small greenhouse, because vegetables will be very expensive in the spring and fall.

How deep are your water pipes? They were put in when the climate was warm. It is no longer warm, and they should be deeper. In the winter of 2013-14 water lines in Manitoba froze at 8 feet, and Winnipeg had 1300 homes without water. Visit your cellar, and add insulation to everything...walls, pipes, hot water tank. If your house is old, consider storm windows, or plastic on the inside, and for North facing windows, consider an insulating plug of foam that can be pushed in for those really cold days.

Imagine what life would be like if you could not get out of your house for a week, as just happened in Boston. Would you have enough basic food, medicine, water? Yes water, because city water mains may freeze. Electricity will go off, and having backup wood or propane heat will be an essential if it is minus 20. Get a medium sized wood stove installed, either in the basement or main floor. Of course melting snow for water in more rural areas is standard practice for 'off the grid' houses, but city air is not the cleanest. A first snow fall in the city should be avoided for water, but one coming right after is falling through clean air, so making water from that would be much cleaner.

Organize your neighbors for community help. There will be elderly folks, the disabled and young mothers who will need backup. Suggest that your local fire department invest in a snow mobile type of vehicle to get people to the hospital.

If you stockpile frozen food, leave the freezer in an outer, but enclosed porch or shed, that way nothing will spoil if the power goes off. Do not buy prepared food, but rather stock up on rice, pasta, canned goods, nuts, grains, dried fruit, powered milk, and canned fish. No sense suffering too much, so stockpile some sugar or honey, and of course chocolate and booze. All these items will disappear from store shelves in the fall, at the first sign of cold weather, and they will keep for years. In Vancouver, a few years ago, un-boiled tap water was unsafe for a week or so because of excessive rain, and fights broke out at Costco over bottled water.

You can rotate food every few years. Eat up your supply and replace it, or better yet, buy a new item, *then* eat the old one. This is like the 'earthquake' kits some people keep in a hall closet on the west coast. We are told: 'its not 'if', its 'when,' but 'when' could be in 20 years. You must assume you will need your food stash <u>next</u> winter.

Check out some of the "Survivalist" and "Prepper" websites. These folks have been preparing for disaster…any disaster for years, and have lots of tips on how to survive. Some of it is geared to Urban Posers but there still are nuggets of good advice on all of these web sites. [114]

After the first few years, re-evaluate and perhaps start stockpiling food to last your family for one or two months. There are only seven missed meals between society and anarchy.

For those who are more adventurous, and want to take this opportunity to move out of the city, there is lots of peace and quiet with some drawbacks to living in the country. The only Appendix to this book offers some advice about making the big move and going back to the land.

During the "year without a summer," 1816" many Canadians and Americans packed up and headed west.

"The grain-growing regions out in the American frontier—present-day Illinois, Indiana and Ohio—actually had decent harvests and the amount of grain they were producing and selling quintupled because of the demand. People made their fortunes. Entire farming villages in eastern North America, starving and weakened, picked up and moved west. Up to that point, it was the largest wave of pioneers settling the vast frontier." [115]

In the event of a major volcanic eruption during this Little Ice Age, it would be advisable to stock up on food, lots of food, because the store shelves will empty after the TV merchants of fear whip up the populace. (I am sure you have seen how effective they were in spreading doom and gloom over 'global warming', the chicken flu, the swine flu, terrorist attacks, Mad Cow disease, and the hole in the ozone layer).

During any big disruption of any kind, the stock markets shudder. I personally will not put any money into the U.S. stock markets, but you might be a bit braver, or maybe you have more confidence in the stock markets in other countries. You now have knowledge of a really big change that is coming for sure. How can you profit from this? Here are a few ideas that might help you acquire some financial security.

The cold is hard on cars. One extra month of cold will cause car parts to quit sooner. The prolonged winters will stimulate a demand for building supplies, both for the do-it-yourselfers, and for professionals. Insulation, lumber, thermal windows, will be in demand.

Learn about which companies make auto parts, and building supplies. Find out who delivers them, who finances them, and which big retailers sell them. Find some investment advisor that is interested in true investing, not just selling you something at tax savings time.

How about stock in holiday airlines and support companies? I know I will be flying South more often.

Grain crops will fail, as will citrus and other fruit crops. There are well established futures markets for these, and many other products, but beware, futures are very risky, because you have to buy contracts for delivery of these products at a future date. You are betting on a future price higher than the price of the contract you bought. Nothing is a sure thing, even bad weather, and there are a lot of factors that can change the value of crops in the field, for example a sudden glut of a product from an unexpected source. Start now getting some advice about this sort of investment.

Land, and buildings in the right place, at the right time could be a very profitable investment. I have bought and sold land and buildings in Canada for 40 years, and never lost money, and I have found that net migration of people is the number one factor in rising real estate prices.

When people flood in, prices go up. Here is a link that you should watch, because Coastal British Columbia is in the Warm Zone, and we will have lots of in-migration. [116]

Net migration statistics exist for all areas of North America, find the web site for yours, or for the area you want to move to.

I have advocated using wood heat either as primary heat, or as backup, because it is the one kind of heat you can count on. No OPEC to raise the price, no truck lines to fail to deliver, no overhead lines to come down and it cannot be taxed. The U.S. EPA (Environmental Protection Agency) has tried to regulate wood stoves to death with new rules, but my guess is that people who are cold will ignore them. Here is a video on keeping warm with wood. [117]

In Germany in 2014, there was an increase in wood pellet furnaces by approx. 36,500 new stoves and heaters to a total of 358,000 units. For 2015, the German Energy Wood and Pellet Association's (DEPV) predicts a 10% percent increase in domestic pellet production.

North America has many wood stove manufacturers, who will probably increase market share in the cold years to come. Those companies, and the manufacturers of things like insulated pipe and other fittings are sure to prosper. This winter the wood pellet manufacturers in Maine ran low and had to ration them. More than 200,000 stoves were sold in the last three years.

The U.S. North East, parts of the Midwest and most of Canada will be very hard places to live in for the next 50 years. But what about the West Coast, the South West, and the South? During the last Little Ice Age in the 1800's it was warmer in the West than it was during the 1900's.

Right now there are houses to buy and rent, or flip in these areas. Markets in some areas may still be depressed after the Bankster meltdown of 2008. Check with realtors in the area you are interested in. Check weather history to see if they get very cold or snowy winters. It is always best to live near your real estate investments. Do not buy bare land, you will need an income from the property to pay taxes so your investment does not eat you up.

Your ability to profit from this coming disaster is limited only by your imagination and bank account. Be sure to diversify, buy stocks such as mutual funds that allow you to hold small amounts of many companies, to spread the risk. Ask around of your friends and people at work to find a broker who is knowledgeable about the stock, bond and futures markets.

There is a shark in every pond, however, so deal with reputable dealers, do not go for the super high returns, and be cautious. If it seems too good to be true, it is. You do not want to get in on the ground floor, you want to deal with well established companies. Avoid the Over the Counter (OTC) stocks, and penny markets, avoid any stock that has to do with "green" technologies. Only deal with brokers who can show you a winning track record, and give them a chance to prove themselves with a small investment for starters. Do not invest in anything new or radical. Those folks unlucky enough to invest in the new medical marijuana businesses will mostly lose their money. The same is true of new technologies… sure, they all say: "get in on the ground floor, we are the new Apple," but for every Apple, there were thousands of pumpkins.

My niece invests in the stock market, and downturns do not bother her at all, she just looks at low prices as a "sale." But, the U.S. stock markets are grossly over hyped, and any investments should await a correction of at least 25%. Keep your eye on the fact that $US Three Trillion in funny money has been injected into the U.S. economy through QE 1, QE 2, and QE 3. (this is quantitative easing, also known as printing money.)

There are also tangible investments such as precious metals and gems, including gold, silver, platinum and diamonds. But here is the catch. Many investment companies want you to speculate or invest in "paper" gold or silver with the promise that you can get the real thing later. But when times get really tough everyone will want delivery of their gold or silver, and there will not be enough to go around. Take physical delivery of the tangible items you invest in, and rent a safety deposit box. Gold at today's prices (about $1200 per oz.) is fairly safe, I think, but there have been news stories about some brokers going 'short' on 'paper' gold. Better ask someone knowledgeable about this.

The cold can be survived, people who were prepared survived the last two Little Ice Ages, and today the Inuit do it every winter. Just like the last Little Ice Age, the people with the most money will survive, because they can buy food and fuel at any cost, so any strategy to get more money is (including gold and silver) is a good idea. (The currency of most countries, and especially U.S. currency is only as valuable as you and others agree it is. It is called "fiat" currency, because it is issued without any commodity, gold or silver back up, and could crash.)

Finally, if all else fails, move. Most of the Western States, and Canada, West of the Rockies there are Warm Zones. The distorted Jet Stream pushes cold air as far south as Mexico, but bypasses the West. The other group of people who will survive will be the self-reliant ones, who have family and community support and who can grow some food, and do with a little less of everything.

Here is a typical winter weather map. Based on temperatures reported on pages 29-32 herein.

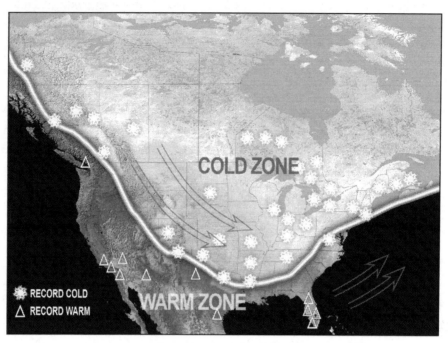

Jet Stream Distortion 2014- 2015

It is possible that the Chinese are preparing for a big move. [118]They have been building 'ghost cities' mostly near the coast in Southern China, but more recently in Africa. In both places these cities are complete: high and low rise apartments, suburban neighbourhoods, lavish government buildings, schools, shopping malls, roads and traffic and street lights, but no people. [119] These photos are staggering. [120]

No one seems to know why they are building these cities for millions of ghost occupants, however a new theory has emerged. It is related to this little ice age. Especially when one considers all the factors, such as food production and cold weather in Northern China, [121]the building of cities in the warmer South, plus the acquisition of farm land in Africa with cities for thousands suggests a deliberate plan to keep millions of people fed and housed in these difficult times. [122] There may be another explanation for building enough housing units for 40 million people, and not filling them up, but I leave it to you to discover it.

"With little fanfare, a staggering 750,000 Chinese have settled in Africa over the past decade. And more are believed to be on their way.

The strategy has been carefully devised by officials in Beijing, where one expert has estimated that China will eventually need to send 300million people to Africa to solve the problems of over-population and pollution." [123] Given China's total population of 1.3 Billion this year, 300 million is a drop in the bucket....unless they are hungry.

The weather in Northern China is as bad in April 2015, with rare snowfall, which damaged or destroyed 5000 acres of wheat, as it has been in Canada and Eastern US all winter. [124]

Some people will be ready to move out of the cities, to the country to avoid the general uncertainty that comes with major transportation shut downs, social unrest and potential food riots.

I was once a subscriber to the "Back to the Land" movement, and left California for rural British Columbia in 1970, just in time for the worst winter in 70 years. [125]

I lived in a drafty shack near Golden, at 3000 feet in the Rockies, and had no electricity, burned wood, and managed to stay warm during minus 40 F winters. My car froze solid, the barn roof caved in, my ears got frost bite (I was used to California weather after all), stuck a chain saw in my leg, and was swatting mosquitos in April, before all the snow had melted. When it did melt, I had a muddy road to navigate for a month or so, and five months later, it started snowing again.

In the "Back to the Land Appendix" at the end of this book there are some ideas for living in the country, especially in the cold,(but do not go up into the mountains, like I did). Before you jump, look at all the costs, especially the hidden ones.

5. Who concealed this coming disaster and what must be done about it.

Several years ago, I was a believer in AGW, (man-made global warming), but I became a bit skeptical when I read a little article, back in 2007 in the newspaper about the climate on Mars getting warmer. [126]

This seemed odd, because I did not believe the Martians had SUVs, and I started doing my own research rather than relying on the Legacy Media (which I also believed in back then). After reading books by Ian Plimer and Lawrence Solomon (both cited herein) my belief in the stories put out by the U.N. IPCC and Al Gore were shattered.

What I do not understand is why the Governments of the U.S. and the E.U. continue with these deceptions, even after all the scientific warnings, these last two cold and snowy winters, and the acknowledgement by former global warming advocates, like the U.K. Met Office that there has been no warming for 18 years. [127]

When I was practicing law in Vancouver, it was well known in the community of personal injury lawyers that some doctors hired by insurance companies were hired because their opinion about the cause of the plaintiff's injury or pain, or the severity of that pain was a foregone conclusion. I have seen doctors become insurance company experts precisely because they always said the same thing.

At least in that setting, we had the opportunity to cross-examine the expert, but the people of the world who were presented with the IPCC's conclusions never had any opportunity to cross-examine that body, or the 'experts' who were part of it, until It was finally examined in detail by a Canadian, Donna LaFramboise.

Governments relied on the IPCC when making enormous decisions concerning the spending of billions of dollars of tax money, allocation of scarce energy resources, and personal freedoms, which ultimately affected the lives and wellbeing of millions of people in the developed world. Now people in the developing world will starve because no one warned them of the coming cold. All those billions of dollars were spent preparing for a warmer world, rather than a colder world.

There was no provable warming. Ian Plimer, the geologist says:

"In the time since the IPCC was formed 20 years ago there has been no demonstration of human-induced global warming....The models suggested constant warming until the end of time. However, neither the post-1998 cooling nor El Nino events were predicted. This alone shows that the computer models are only sophisticated computer games with input based on the programmer's predilection. Climate predictions are not evidence and certainly are not suitable for environmental or political planning. [128]

The green corporations obtain massive amounts of money from foundations and memberships every year. In 2012 Greenpeace USA was the recipient of $32,791,149 and that is true of other environmental pressure groups. In 2012 $111,915,138 was secured for the Environmental Defense Fund, $98,701,707 for the Natural Resources Defense Council, $97,757,678 for the Sierra Club, and, for Al Gore's Alliance for Climate Protection, $19,150,215. [129]

In addition to putting out misleading stories about the climate, Greenpeace in particular, along with others are attempting to smear legitimate scientists who have expressed doubt about the alarmist's claims. Science is about doubt and skepticism keeps things honest. [130]

Science is supposed to test theories by publishing, then waiting for feedback; positive or negative, so as to advance the understanding of a particular topic. However, in the world of climate science, dissent is not tolerated, because at its core, global warming is a fraud. No dissent can be allowed, and the alarmists with their green corporation cash are trying, in the dying days of this hoax, to discredit anyone who disagrees with them, thus bringing in 21st Century fascism. [131]

The Climate-gate emails, leaked in 2009 and 2011 reveal a concerted effort by those in the 'global warming' camp to stifle dissent by attacking editors of science journals that disagreed with them, finding friendly reviewers for their work, and working together to circumvent freedom of information rules. [132]

We have been presented with many inaccuracies and half-truths by our Governments the U.N. Intergovernmental Panel on Climate Change (IPCC), by green corporations like Greenpeace, and the WWF, and 'scientists' who will say anything for a fee, all echoed by the Legacy Media.

One of the biggest inaccuracies put forward by these groups and others is that the Greenland Ice Sheet is melting or is about to melt, flooding the world, because of "carbon pollution." (NOTE: this refers to carbon dioxide, the gas that we exhale and trees absorb and an essential ingredient for life on this planet. CO2 is also referred to as "emissions.") The other is about the disastrous melting going on in Antarctica.

Here is what a few alarmist web sites have said over the last few years:

Greenpeace:

"Greenland's glaciers are melting even faster than previously thought and contributing more and more to sea level rise caused by global warming. If you live near the sea and think global warming isn't a problem for you, it's probably time to think again." [133]

Live Science:

"...scientists might be underestimating Greenland's ice loss" [134]

Greenpeace:

"Greenland and Antarctic ice sheet melting. Unless checked, warming from emissions may trigger the irreversible meltdown of the Greenland ice sheet in the coming decades, which would add up to seven meters of sea-level rise, over some centuries; there is new evidence that the rate of ice discharge from parts of the Antarctic mean that it is also at risk of meltdown." [135]

Planet Extinction:

"It would seem we are on the verge of a major tipping point (in Greenland) in climate

change" [136]

Climate Central:

"The increased melt (in Greenland) raises grave concerns that sea level rise could accelerate even faster than projected, threatening even more coastal communities worldwide." [137]

Greenpeace:

"In the case of Greenland, loss of between 50 and 100 billion tons of ice has taken place annually over the period 1993-2003 with evidence of higher rates more recently. Recent research has suggested that the ice sheet could melt completely over the very long term (millennia) if global temperatures rise by somewhere between 1.9 and 4.6 degrees and they have already risen by an estimated 0.8 degrees. Complete melting would lead to a sea level rise of around 6-7m." [138]

Many measurements dispute the alarmist's claims, but these two stories make a mockery of the alarmism.

The most compelling story about the increasing ice on the Greenland Ice Sheet is one from 1942, during the Second World War. A squadron of six P-38s Lightning fighters and two B-17 bombers was flying to England from Greenland when they got lost in the fog, and ran low on fuel but were able to make a wheels up crash landing on the ice. The crew was rescued, and most people forgot about the airplanes which were abandoned because of the cost to recover them.

As time went on however those brand new airplanes would bring quite a price from collectors, so several teams went looking for them. Despite knowing of the exact location of the crash site, they were nowhere to be found. Finally, in 1989 after bringing in a special icescope which could detect metal below the ice, all the airplanes were found under *286 feet of ice, or 27 stories.* That was 6 feet of ice per year for 47 years. [139]

One of the P38s was hauled up out of her icy tomb, refurbished and now does the air show circuit under the 'Nom de Show': "Glacier Girl." [140]

A documentary film, made 1990-2000 about the locating, recovery and rebuilding of "Glacier Girl" reveals the incredible snow that falls on the Greenland Ice Sheet in the summer, burying the explorer's tents daily. [141]

A similar story concerns the bases built in Antarctica over the years. The buildings are being buried by the ice. The original Byrd station, built in 1957, is now buried beneath 45 feet of ice. Glaciers are advancing at twice the normal rate, and five of the six which feed the Larsen Ice Shelf are surging. [142]

The alarming story about the West Antarctic ice sheet collapsing which was front and center in the Lame Stream Media last year sounds serious, until one realizes that the only melting going on there is because of undersea volcanos. Antarctic ice has been growing.[143]

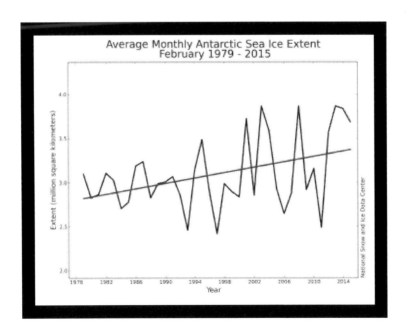

Graph from the National Snow and Ice Data Center. [144]

The U.N. IPCC is the original source of this global warming hysteria. It was created in 1988, and was seriously beefed up after the Rio Environment Summit of 1993. It is not a scientific body, but rather a political one, but holds itself out as:

"The IPCC is a scientific body under the auspices of the United Nations (UN). It reviews and assesses the most recent scientific, technical and socio-economic information produced worldwide relevant to the understanding of climate change." [145]

The IPCC, in conjunction with a compliant media, and reporters and politicians who were prepared to believe what they said, launched the great global warming hoax. Many believe 'global warming' was to provide funding through the Kyoto Protocol, fines and penalties, commissions on cap and trade schemes, and payment of 'climate reparations." The goal of all of this was to become a self- financing organization that would later morph into an EU style (or worse) global government, run by unelected bureaucrats.

Canada withdrew from Kyoto in 2011, and avoided $14 Billion in penalties. It was a contentious decision, but clearly the right one. Our representatives at various climate meetings and gab-fests had to endure the name calling associated with our position. Canada was awarded two awards, the "Fossil of the Day" and the "Rio Fossil," which I am sure will be something for all Canadians to be very proud of in the not too distant future. [146]

When the IPCC is examined in detail it is, I believe, a propaganda organization created solely for some other purpose, including possibly the purpose of financing some greater project. The IPCC went wrong from the very beginning, starting with the terms of reference. [147]

The Terms state that it is "to provide the governments with objective, comprehensive and up-to-date information about *climate change and its implications*." (Emphasis added) Hence the name: The Intergovernmental Panel on Climate Change...not the Intergovernmental Panel on Global Warming," but that is what it became.

It was hijacked at the beginning for someone's purposes. I believe we can just 'follow the money' to find out what the whole purpose of the scheme was. Given the looming disaster, an international body should undertake this work, and prosecute those responsible.

Donna LaFramboise, a Canadian writer and investigative journalist has published an expose of the IPCC and its (former) chairman, Rajendra Pachauria. "The Delinquent Teenager who was mistaken for the world's Top Climate Expert" has been praised by many, including Richard S.J. Tol, Professor of the Economics of Climate Change and convening lead author of the IPCC. Ms. LaFramboise was invited to give evidence to the UK Parliament, House of Commons Energy and Climate Change Committee in 2014 about her findings.

She was the first person to 'cross examine' the IPCC by doing what anyone could do: look at what they said, what they did, and how they did it to find the truth. Her first major discovery concerns the organizational structure of the IPCC, and how the 'independent' researchers who contribute to the reports are funded.

Because it is a U.N. body, great attention is paid to "International, political considerations…. "(w)hile scientific credibility comes second." [148]

This means that the people contributing to the IPCC reports are first and foremost representing the interests and biases of their countries. "The IPCC has never asked the science academies of the world to identify the top experts in a number of fields-and then systematically recruited only those people. Instead nominations are sought from governments." Those nominees come loaded with the political baggage of their governments. [149]

In Europe, the Green Party is often awarded the environment portfolio, and a Green nominee would have much different baggage than a Conservative nominee.

Let us not forget that the 'third world' has consistently clamored for 'damages and reparations' to be paid by the 'first world' for 'climate crimes.' Robert Mugabe, one of Africa's longest serving tyrants went on a rampage at one of the recent climate gab-fests demanding billions of dollars. I wonder how the Zimbabwe delegation handled the question of global warming? What would have happened to those delegates when they got home, if they questioned the warming narrative? Total membership in the U.N. is about 195 nations, of which the First World is about 25.

The Kyoto treaty promised that the First World would pay the U.N. for failing to meet climate targets, and the resulting cash flow would have been administered by the U.N. for a fee. The surplus would have been divided up among the Third World.

Governments allocate money for research into all sorts of things, but administrators dole out the money. Any good bureaucrat knows that the path to power and success is to expand their budgets, and ringing the climate alarm bells gets much more attention and funding. [150] In much the same way as the Legacy media focuses on disasters: "if it bleeds, it leads."

Aside from the method of choosing participants, once chosen, the IPCC has conducted its business in a way that would have guaranteed bankruptcy or jail in the free market.

The now disgraced chairman, Rajendra Pachauria told the world, and politicians many times, sometimes under oath that the contributors to the IPCC reports were at the top of their professions, the best in the world, and that they only relied on peer-reviewed scientific papers. [151]

However the truth is that they recruited inexperienced people, listened to environmental activists, relied on press releases and allowed incestuous relationships to grow both among individuals and with publishers of scientific papers. [152]

A 'citizen audit' of the thousands of references to scientific papers found in the IPCC reports, revealed that 30% were not peer-reviewed. "Among the sources used to support IPCC assertions were newspaper and magazine articles, unpublished Masters and doctoral theses. Greenpeace and World Wildlife Fund Documents, and yes, press releases." [153]

An example of these problems is best illustrated with Mr. Bill Hare: "(he) has been a Greenpeace spokesperson since 1992 and served as its 'chief climate negotiator' in 2007. A Greenpeace blog post describes him as a legend in that organization." "When the 2007 (report) was released, we learned that Hare had served as a lead author, that he'd been an expert reviewer for two out of three sections of the report." [154]

But he was not the only Greenpeace disciple who got into a position of importance.

"Richard Klein, now a Dutch geography professor, is another example. In 1992 Klein turned 23, completed a Masters degree, and worked as a Greenpeace campaigner. Two years later, at the tender age of 25, he found himself serving as an IPCC lead author." [155]

There was no quality control, no checking of the facts that conclusions were based on, and if outsiders asked for the data, it was sometimes denied. [156]

An excellent example of not checking facts was the graph that Dr. Michael Mann showed up with one day. It has come to be known as the "Hockey Stick" graph of global temperatures. It had a long flat handle, and at the end a sharp upturn reminiscent of the blade of a hockey stick. It purported to show world temperatures from 1,000 to present as being flat, with the sudden steep increase just recently. Humanity was to blame. The industrial revolution and the burning of fossil fuels, which have brought prosperity, longer life spans and security to millions was the villain.

To create such a graph of world temperature it would be necessary to eliminate the Medieval Warm Period, and the Little Ice Age which would have added 'bumps' to the handle, one up and the other down. Both of these periods of time had been studied extensively, by hundreds of scientists and their papers were peer-reviewed.

The 'blade' purported to show a sharp up-swing in temperatures since 1850 or so, and was blamed on industrialization and the 'green house' effect of carbon dioxide. It quickly became "Exhibit A" in the global warming hysteria. It was reproduced in several IPCC Reports, until even they became embarrassed and threw Dr. Mann under the bus. He has been fighting lawsuits ever since. [157]

To include this data, offered by a freshly minted PhD, should have required extensive investigation before overturning the work that had taken hundreds of scientists decades to compile.

Two Canadians, one of whom was familiar with mining stock scams recognized the "Hockey Stick" as the sort of graph that mining companies put out to drum up business for their stock. The pitch for the miners as well as the alarmists is: 'look where we are going,' [158]

They requested Dr. Mann's computer code, to do what normal science does, try to reproduce his results. Dr. Mann resisted producing his code for years, all the while the media had a frenzy with the coming global warming meltdown, Al Gore promoted his famous movie: "An Inconvenient Truth" and millions of children grew up believing in the 'greenhouse effect' and dangerous 'global warming' before Mann produced the code. But by then the damage had been done.

Mcintyre and McKitrick wrote a report on Dr. Mann's calculations, and the data that was processed with his computer code. In one cross check of his work they found that random numbers generated a 'hockey stick' shape, and in the majority of simulations the same shape was produced, either with the blade pointing up, or down. His work was considered by them to be inaccurate. [159]

Just recently, I heard Bill Nye, the children's television entertainer, billed as "The Science Guy" explaining global warming, and he referred to the Hockey Stick, as though it was accurate and helpful.

The IPCC broke its own rules. There was a requirement that data such as press releases should be identified, however on examination only 0.1% were flagged, and when these rules were questioned, they simply changed them, with no consequences. [160]

The reports, on major issues, "systematically conceal or minimize what appear to be fundamental scientific uncertainties." When climate computer modelers (who contributed much to the reports) found inconsistencies between real world observations and their models, they assumed the observations were wrong. [161]

But what is worse is that there were also assumptions made about feedback loops. In natural systems, if, for example a gas is released into the atmosphere, it will have a high concentration at first, but will then dissipate. This is a 'negative feedback.' An assumption that there was a positive feedback would see the concentration of the gas increase.

The Climate modelers assumed that contrary to most other natural systems CO2 in the atmosphere would increase.

"Incredible as it sounds, therefore, the only reason climate models tell us we are at risk of eco apocalypse is because the climate modelers believe our climate system behaves in a manner that is opposite to the way most natural systems behave." [162]

"It is both peculiar and ironic that an organization that so vigorously claims to represent a worldwide scientific consensus has systematically 'disappeared' so many consensus views held by so many different kinds of researchers. The IPCC ignores the consensus among hurricane experts that there is no discernible link to global warming. It ignores the consensus among those who study natural disasters that there is no relationship between human greenhouse gas emissions and the rising cost of these disasters. It ignores the consensus among bona fide malaria experts that global warming has not caused malaria to spread.

In each case the IPCC substitutes its own version of reality. In each case that version of reality makes global warming appear more frightening than genuine experts believe the available evidence indicates." [163]

The global warming hoax was never about our climate, or saving the earth and all the little creatures, but was about wealth re-distribution, and money for green corporations, green profiteers and the U.N. There is even now an admission by a top U.N. official that the goal was the transformation of the world economy. [164]

Again, I ask the question: "how did they all miss the descent into a new Little Ice Age"? I and others have found dozens of scientific peer-reviewed papers on this topic, and on related and intersecting topics yet the people who were supposed to advise world governments about "climate change" missed it. How did they miss it? Why the global warming tunnel vision? Who was in charge? Who is going to pay the price for this stupidity?

The answer to at least one of these questions: 'how did they miss it,' may be found in Dr. Tim Ball's new book. Dr. Ball is one of the scientific leaders in climatology who dismisses the idea of a global warming 'hoax.' He says quite clearly that it was a very deliberate deception rather than a 'hoax.' If that is the case, then maybe they did not miss the potential for cold weather killing people, especially if we were expecting warmer weather. Deliberately causing deaths is a far different matter from standing idly by while they occur. Furthermore, it is not a crime under Canadian law to help spread false stories that may result in death. However I believe that the International Criminal Tribunal for Rwanda did convict at least one person for stirring up the masses with false stories, leading to the deaths of 800,000 people. [165]

"The Deliberate Corruption of Climate Science," published in 2014, Ball says that this was a political act by a minority who were connected by ideology. They perverted the scientific method, silenced critics, and enriched themselves with grant and research money. These people, in his opinion had a hard Left agenda.

Part of that agenda, at least for John Holdren, who is now Obama's science advisor, was to reduce the population of the Earth. In a paper he wrote many years before being nominated to his present position, he advocated: forced abortions, large scale sterilizations via chemicals in the drinking water, seizure of babies from single mothers, and creation of a transnational "Planetary Regime" which could assume control of the global economy and dictate the most intimate details of American's lives, backed up by an international police force. [166]

After he got the job, his previous comments came to light, but Mr. Holdren said he no longer believed in those things, however looking at the Obama White House's record, one might question that statement. The global warming hoax and the U.N. treaty banning weapons, as well as the dispute over abortion funding in Obamacare suggest the opposite.

John Holdren made predictions about global warming and other matters of science that turned out to be grossly wrong. He worked with Paul Ehrlich, the author of the completely inaccurate "The Population Bomb" and together they advanced the idea of limiting or reducing population.[167]

Even though Mr. Holdren says he does not hold those views now, what other views about the lives of ordinary people might he hold? Can anyone who ever put such radical views in writing ever be trusted? Should such a radical ever be given such an important post?

Dr. Ball discusses the deliberate actions within the IPCC that he believes proves that the entire 'global warming' idea was a falsehood from the beginning, and the steps taken to sell it to an unsuspecting public.

Those actions include issuing the "Summary for Policymakers" (SPM) for each new Science Report (SR) before actually issuing the SR chapter. Since the chapters, written by scientific committees often offered other possible conclusion than alarmism. "...the...SPM, the most important part of the IPCC work, which is always dramatically different than the Science Report," was released months before the Science Report, to give the media time to spread the alarmism. [168]

The Science Report would, of course, be the 'science' part, but the Summary for Policymakers was the political component, and that is what was used to help us form opinions about 'global warming.' "It is very likely that few people ever read the Science Report. It is more logical to produce the Science Report first and then the SPM. There is only one explanation for producing it first. The final product achieved the result of deception in full daylight." [169]

Many other scientists who wrote papers critical of the global warming hysteria were sidelined, fired, discredited and had trouble getting climate journals to publish their papers. [170]

The U.N. has demonstrated extreme incompetence and bureaucratic rigidity in the past. In 1994 the Rwandan genocide happened. Lt. General Romeo Dallaire, the Canadian leading the U.N. peace keeping troops in Rwanda recounts his maddening attempts to act to prevent the slaughter. Even though knowing of the location of weapons and the intentions of the parties, he was not given permission to act by U.N. bureaucrats.

He wrote:

"The UN must undergo a renaissance if it is to be involved in conflict resolution. This is not limited to the Secretariat, its administration and bureaucrats, but must encompass the member nations, who need to rethink their roles and commit to a renewal of purpose. Otherwise the hope that we will ever truly enter an age of humanity will die as the UN continues to decline into irrelevance." [171]

Both the governments of Australia and Canada have rejected carbon pricing (which is promoted as a means of controlling Carbon Dioxide emissions, but really is just another wealth redistribution scheme) because, in the words of Tony Abbott, Australia's Premier: climate change is "absolute crap." [172]

Canada's Prime Minister has publicly stated that Canada stands with Australia on carbon taxes. He is way too conservative to utter the word 'crap' in public, but he should be right up front with Canadians about the coming Little Ice Age. He talks a lot about protecting us from criminals, but what about the cold? [173]

There have already been casualties in the U.K. from the disastrous "green" energy policies adopted by their government. 20,000 seniors died from the cold, or cold related illnesses in 2011 because of 'fuel poverty,' which has been called 'the deaths that shame Britain.' A spokesman for 'Age UK' said: "Fuel poverty is defined as when a household needs to spend 10pc or more of its income on maintaining an acceptable level of heating throughout the whole property. In England, according to the latest official figures, there are 3,964,000 households in fuel poverty, over half of which contain someone aged 60 or over as the oldest person in the household." [174]

I find it amazing that some governments have forgotten that a big increase in food prices often signals a change in government. It happened in the last Little Ice Age, during the French Revolution. That was some change!

Some people more cynical than I am, have suggested that the highest levels of government in the U.S. knew when they printed trillions of dollars, drove up the national debt to over $17 trillion (total U.S. debt of $60 trillion) and destabilized the Middle East, that after the Great Starving of the 2020s, during the Eddy Minimum, (as the current approaching Minimum has been named) [175] there would be either no one to demand repayment, or resist any NATO/ US foreign adventures. Is it any wonder that Vladimir Putin is flexing his country's military muscles?

Russia possesses the world's expert on solar cycles and the coming new little ice age, Dr. Habibullo Abdussamatov, so it is reasonable to assume that global warming hysteria is not a factor in decision making in Russia, which is bad news for the U.S., and her allies.

Napoleon withdrew from Russia in the very cold winter of 1812. He had marched in with 680,000 men, but by the time he got home there were 50,000 left. Tyrants sometimes have trouble with the weather.

Your local, state and provincial governments, as well as federal governments in the US and Canada have "Environmental" departments and "climate change" bureaucracies and they spend lots of your money on "research" on "global warming" issues, and planning for 'mitigation' of global warming, and 'sustainable' solutions for a 'warming world.'

It is time for heads to roll, and grants to be cut. Encourage governments to start spending your tax dollars on helping people with fuel costs and home improvements. Just think, all those "Environment Researchers" and global warming 'scientists' completely missed the coming Little Ice Age, and now they expect you to suffer while they continue getting their paycheck.

Any grant receiving 'scientist' who offered opinions concerning any aspect of the 'global warming' hoax, who did not at least suggest that there was a possibility of a major cooling trend, should be cut off for life from taxpayer money, or perhaps for a shorter period if some sort of adjudication finds that a lesser time out would be appropriate. After all, Robert Kennedy Jr. wanted to jail people like me. [176]

Others have called for a modern equivalent of witch burning, including Al Gore who said: "We need to put a price on carbon to accelerate these market trends, and in order to do that, we need to put a price on denial in politics." [177]

I said when I was practicing law that carelessness, stupidity and greed kept me in business. With the 'global warming' 'scientists' it is all three. In a normal free market economy those qualities are not rewarded.

The Australian Government recently fired the entire board of CSIRO, the agency in charge of climate change issues. [178]

Global temperature data has been falsified, by using models rather than actual recorded temperatures, because the actual temperatures either showed cooling, or no warming and did not fit the narrative of global warming. [179]

Ocean acidification, the ugly stepsister of global warming appears to have been invented, and data manipulated to show trends that did not exist. A whole industry of 'research' into the effects of ocean acidification was created, with each researcher boot strapping herself up on the previous shoddy research, and at the end of each research paper, a plea for more research, ' to better understand what is happening.' The big scare was that the oceans would become so acid because of the burning of fossil fuels that shells on clams and other shell fish would dissolve. [180]

Ian Plimer, a geologist, says that based on millions of years of recurring high levels of CO2 in the atmosphere, the fossil record continues to record fossils with shells, so they did not dissolve in the past. Furthermore, the addition of more CO2 would only make more calcium carbonate (CaCO3, better known as limestone) and there is layer of limestone 4.8 kilometers deep in the ocean. "Acidification is greatly exaggerated in the popular media as a potential environmental catastrophe." [181]

Many research papers say one thing, but the 'abstract,' or summary of the study says something else, usually that 'global warming' caused the problem. This is the same tactic used by the IPCC when releasing the Summary for Policy Makers that said one thing, while the Science Report said another. The truth was in plain sight, but one had to read the whole document to see it.

There have been hundreds of 'problems' that have been blamed on 'global warming.' This website has a complete list (complete at least until someone thinks up a new 'problem' to blame on global warming).Such matters as animals shrinking, birds not coming to the UK, the Earth slowing down, growth in ice sheets and potential conflict with Russia are, according to some researchers to be blamed on global warming. There are hundreds more at this site. [182]

Climate-gate

In 2009 and 2011 large numbers of emails and documents from the Climatic Research Unit (CRU) of East Anglia University were released to the public by persons unknown. The CRU was and is one of the worlds best known institutions for the study of the climate, and a regular contributor to the IPCC. These leaked e-mails were the first crack in the armor of the global warming hoax. Writing in <u>Forbes Magazine</u> in 2011, James Taylor says:

"Three themes are emerging from the newly released emails: (1) prominent scientists central to the global warming debate are taking measures to conceal rather than disseminate underlying data and discussions; (2) these scientists view global warming as a political "cause" rather than a balanced scientific inquiry and (3) many of these scientists frankly admit to each other that much of the science is weak and dependent on deliberate manipulation of facts and data.

Regarding scientific transparency, a defining characteristic of science is the open sharing of scientific data, theories and procedures so that independent parties, and especially skeptics of a particular theory or hypothesis, can replicate and validate asserted experiments or observations. Emails between Climategate scientists, however, show a concerted effort to hide rather than disseminate underlying evidence and procedures." [183]

We know from the 'Climate-gate' e-mails that editors of climate related publications were under pressure, and the peer-review process had been perverted. [184]

Prominent scientists such as Ian Plimer are crucified by the Warmist Alarmists.
Jonathan Manthorp, writing in the "Vancouver Sun" in 2009 said:
"Ian Plimer has outraged the ayatollahs of purist environmentalism, the Torquemadas of the doctrine of global warming, and he seems to relish the damnation they heap on him. Plimer is a geologist, professor of mining geology at Adelaide University, and he may well be Australia's best-known and most notorious academic." [185] He is a major academic figure in Australian geology, and was the two time winner of Australia's highest scientific honor, the Eureka Prize.

Hockey Stick

The infamous "Hockey Stick" graph created by Michael Mann and used by the IPCC and Al Gore to strike fear into the populace was dismantled by a senior mathematician in a report commissioned by the U.S. House of Representatives. Edward J. Wegman and his committee pointed out the statistical errors committed in coming to the conclusion that world temperatures were rising at an alarming rate. He also commented on the cozy relationship among climate 'scientists' who peer-review each other's work.

"In our further exploration of the social network of authorships in temperature reconstruction, we found that at least 43 authors have direct ties to Dr. Mann by virtue of coauthored papers with him. Our findings from this analysis suggest that authors in the area of paleoclimate studies are closely connected and thus 'independent studies' may not be as independent as they might appear on the surface. This committee does not believe that web logs are an appropriate forum for the scientific debate on this issue."

"It is important to note the isolation of the paleoclimate community; even though they rely heavily on statistical methods they do not seem to be interacting with the statistical community. Additionally, we judge that the sharing of research materials, data and results was haphazardly and grudgingly done. In this case we judge that there was too much reliance on peer review, which was not necessarily independent. Moreover, the work has been sufficiently politicized that this community can hardly reassess their public positions without losing credibility. Overall, our committee believes that Mann's assessments that the decade of the 1990s was the hottest decade of the millennium and that 1998 was the hottest year of the millennium cannot be supported by his analysis." [186]

Such a thorough condemnation would, in any other scientific discipline, or most any other endeavor do irreparable damage, but Dr. Mann's mutant theories continue to hold some sway in the popular media.

Fortunately there is a 'silent majority' of people in the science community who are skeptical of the global warming crisis. It is a pity they did not speak up sooner. [187]

Over 5400 Wikipedia articles on global warming were either censored or re-written by William Connolley, a global warming zealot with a PhD, until he finally got caught, and was removed from his editorship with Wikipedia.

"He routinely deleted entries that presented competing views and barred contributors with whom he disagreed. He also smeared scientific skeptics by rewriting their online biographies."[188]

During his time at the keyboard secretly making his own 'consensus,' how many people were influenced by his words, how many people continued to support throwing away billions on 'global warming' and 'climate change' initiatives?

Why do so many people do the bidding of the IPCC and their U.N. masters? Ian Plimer has an answer:

"Human-induced global warming is a popular belief because it offers the satisfaction of righteousness without actually having to do anything. Subscribing costs nothing, it provides the immediate reward of moral superiority, and there is the bonus of seeing "polluters" having to pay for their sins. This makes an attractive package and the media provides an abundance of pseudo-evidence to support it. Trying to counter this with considered thoughtful scientific argument which can't be conveyed in a seven-second sound bite cannot compete in popular appeal." [189]

There will soon be another reason in the U.S. to conform to the lie of man-made global warming. Just as in the Medieval practice of Trial by Fire, where an accused had to walk barefoot on red hot ploughshares. If she was not burned that proved innocence, but injury proved guilt, and she was executed.

The Governors of all states must approve plans, starting in 2016 to deal with mitigation of hazards caused by global warming. If they refuse, the Federal Government may block their states' access to hundreds of millions of dollars in FEMA funds. Over the past five years, the agency has awarded an average $1 billion a year in grants to states and territories for taking steps to mitigate the effects of disasters.

"If a state has a climate denier governor that doesn't want to accept a plan; that would risk mitigation work not getting done because of politics," said Becky Hammer, an attorney with the Natural Resources Defense Council's water program. "The governor would be increasing the risk to citizens in that state" because of his climate beliefs. [190]

Now it is time to get the 'climate leeches' off the public payroll, including the private investors in "green energy". The irony of everything named 'green' is that George Orwell could not have done better. The idea with all these schemes is to reduce carbon dioxide, the gas that makes things green. I guess that if it had been called 'brown energy' it would have flopped sooner than it did.

Here is how the green energy profiteers work: they make a deal with your government to generate solar or wind energy then sell it back for say 15 cents per kilowatt hour (kWh). Then you are charged say 5 cents more per kWh, so each kWh then costs 20 cents. Everyone is happy and feels good, except YOU. Electricity can be generated by burning clean coal for 3 or 4 cents per kWh. You just paid, directly and indirectly 16 cents per kWh more than you should have. All in the name of 'reducing greenhouse gasses' and 'saving the planet.'

Tim Ball says that a U.S. Senate report confirms that for all types of power coming on line by 2015, solar will be 173% more expensive per unit of energy delivered than traditional coal, while wind power is 42% more than nuclear or natural gas. [191]

All "green" power needs backup, an idling spinning reserve, because the sun and wind are not constant. So that power plant that burns coal or oil, cannot be retired; it needs to keep running, all the time, to backup those "green" energy sources. Furthermore, a conventional power plant is not as efficient when idling. You are paying TWICE, and whatever environmental improvement you thought you were buying into is just not there. [192]

Burning coal may release more carbon dioxide than other fuels, and CO2 does warm up the Earth a tiny bit. Since it is getting colder and food will be scarcer, we should burn as much clean coal as we can. Clean coal has been used in the U.S. for years, generating perhaps as much as half of U.S. electricity. It is made clean by first crushing the coal, soaking it in a special liquid to remove impurities, then when it burns taking out the sulphur and oxides of nitrogen with other technologies. Every commercial greenhouse raises the CO2 levels from background of 400 ppm to 1200 ppm to squeeze out as much produce as possible.

Here is a video that very powerfully demonstrates the value of CO2 as plant food by comparing two test crops of cowpeas, one in a 450 ppm CO2 environment, and the other in a 1270 ppm environment. The results are staggering. In the higher concentration, the stem height was 52% longer, the stem weight was up 21%, while root length was 339% longer and root weight was improved by 143%. We could have huge increases in food worldwide with a CO2 level of 1200 ppm or more. [193] Plants stop growing when CO2 is below 200 ppm.

Government health and safety standards set limits for human exposure to CO2. Typically in Canada as in the U.S. average exposure limits are 5,000 ppm, while short term exposure can be as high as 30,000 ppm. [194]

"Massachusetts residents are really getting the shaft from wind power. Thanks to the state's Green Communities Act of 2000, customers may see their energy rates rise from 8 cents to 31.3 cents per kilowatt-hour over the next 15 years." [195]

The U.S. government has lost billions on 'green energy' projects. At least 34 companies are bankrupt or in serious trouble, losing billions of tax dollars in the process. [196]

When a government picks winners and losers, the biggest loser is always the taxpayer, because the free market, devoid of subsidies, cronyism and corruption, will always allow the best technologies to succeed. [197]

'Green energy' is not the only 'green' thing eating up tax dollars by crony capitalists. Obama's green jobs have, as of 2013, lost taxpayers half a billion dollars. Any president could have done it by throwing money into projects with no hope of success, and no 'invisible hand' to make those business decisions. 198

When government colludes with the U.N., reform is urgently needed, because the next scam is 'sustainability and diversity.' Already the poor are feeling the U.N.'s lash in the third world because of the creation of international parks and protected refuges. In the last 10 years these areas have doubled to more than 108,000. [199]

"Crony capitalism occurs when the government interferes with the economy, creating advantages for some businesses at the expense of others. When government colludes with businesses to distort the free market, inefficiencies are introduced into the economy. A popular form of crony capitalism is subsidies to renewable energy companies. Government policies that favour renewable energy companies are common because politicians believe such policies will help fight climate change, reduce dependency on oil, and protect the environment for future generations—all while creating "green jobs." [200]

U.S. expenditures for global warming related initiatives have cost taxpayers hundreds of Billions of dollars, with nothing to show for it, since there has been no warming, and now the taxpayer tap will have to be turned on to fight the New Little Ice Age. [201]

On June 6, 2013, Dr. Roy Spencer said:

"Hundreds of millions of dollars that have gone into the expensive climate modelling enterprise has all but destroyed governmental funding of research into natural sources of climate change. For years the modelers have maintained that there is no such thing as natural climate change…yet they now, ironically, have to invoke natural climate forces to explain why surface warming has essentially stopped in the last 15 years!"

" Forgive me if I sound frustrated, but we scientists who still believe that climate change can also be naturally forced have been virtually cut out of funding and publication by the 'humans-cause-everything-bad-that-happens' juggernaut. The public who funds their work will not stand for their willful blindness much longer." [202]

In Canada, Ontario energy consumers are the biggest losers in the 'green energy' casino. "Ontario residents will pay an average of $285 million more for electricity each year for the next 20 years as a result of subsidies to renewable energy companies." [203]

And now, the Liberal Premier of Ontario, Kathleen Wynne is proposing to bring in a 'cap and trade' system, which will do nothing to improve the economy or environment, but will pour more tax dollars into the Province's treasury. She must have forgotten the proposals by the federal Liberal Party to bring in a similar scheme in 2007. It was attacked by the Conservatives as 'a tax on everything,' and the voters agreed in 2008 by electing a Conservative Government.

The European Union is proving to be the master of wasted funds on 'green' projects.

"In a blog post titled Europe's $100 Billion Green Energy Mistake, Walter Russell Mead draws our attention to a newly-released report about the future of electricity.

The report says that, in their eagerness to embrace renewable energy, European governments have funded solar capacity in countries that get little sun, and installed wind turbines in nations that get little wind.

To quote page 14 of the electricity report, this "suboptimal deployment of resources" has cost the European Union "approximately $100 billion" more than if those governments had taken into account facts that seem "obvious to most European citizens."" [204]

The final organization that needs a complete overhaul is the United Nations. There can be no doubt that they are not with us. (It is hard to divine who they are with.) This organization has, since the 'Oil for Food' fiasco demonstrated that a source of money independent of contributions of nations was desirable. Why? What do they want with a secure independent source of funds? Why do they keep pushing more 'treaties' that take more control of the world? The global warming scam almost gave it to them. The climate change fund passes through U.N. hands, and they are now proposing even more draconian 'taxing' powers. [205]

The U.N. has been after First World money for years, claiming 76 Trillion for green technologies and climate reparations, and later demanding $50 to $100 billion PER YEAR for the third world, for a "climate debt." [206]

But now the battle heats up. Heading to Paris for the climate gab-fest 2015, in November and December, is a new demand on the table: payments for past crimes. This is for sure the U.N.'s Jump the Shark Moment. Our great-great-great-great grandparents committed climate crimes back to 1750, so the U.N. wants us to pay for it. Here is the astonishing wording from the draft points for discussion for the new 'treaty'.

""Each Party to the Convention whose per capita greenhouse gas emissions exceed the global average per capita greenhouse gas emissions" shall be listed as an "Annex I" nation, which means its citizens will be assigned a 300-year "carbon debt" for the period 1750-2050." And called on to pay it. We're not talking mere hundreds of billions of dollars here. As we have reported previously, various UN proposals have demanded tens of trillions of dollars as "climate reparations."

There is to be "Established the International Climate Justice Tribunal in order to oversee, control and sanction the fulfillment of and compliance with obligations of Annex I and Annex II Parties under this agreement and Convention." [207]

Normally governments tax us, and have legal processes to enforce those tax rules. All the rest is just about spending the taxes. It sure sounds like the foundation of a World Government that all the 'conspiracy theorists' have been talking about. Remember when George H.W. Bush spoke of the New World Order headed by a 'credible' UN? The speech is still on Youtube.

The Rio Earth Summit of 1992 was the breeding ground for a new type of world scourge: Agenda 21.

Tim Ball writes :

"Agenda 21, an abbreviation for a plan for the 21st century set out in its Principals a template for achieving global governance and transfer of wealth. Among them was Principle 15 that effectively waives the need for science" [208] by bringing in the 'precautionary principle,' a device loved by those who would stop all development, because it does not require any facts or science. The 'precautionary principle' decrees that if something <u>might</u> be harmful, then it cannot go ahead unless the person proposing it can prove it will <u>not</u> be harmful. How easy is it to prove a negative?

Elizabeth Nickson, columnist, investigative journalist, and novelist, former European bureau chief of Life magazine and a reporter for Time magazine presently lives on Salt Spring Island in B.C. Her book "Eco-Fascists" was the result of dealing with the sorts of bureaucracies that we will all have to deal with if the UN is not reigned in.

"More than 10 percent of the developing world's landmass has been placed under strict conservation—11.75 million square miles, *more than the entire continent of Africa.*" [209]

That 'strict conservation' means that the people who lived on those lands had to either quit doing certain things, like traditional farming, or move. "Millions now live in shantytowns ringing their former homelands in Africa, South America, and Southeast Asia..." [210]

The U.N. has plans for North America which are being delivered by 'soft laws' through organizations such as ICLEI, and through friendly 'green' governments such as the current Democratic Government in the U.S. which is actively tying up land with new water laws (WOTUS) and the creation of National Monuments. "At present 30 percent of the nation's land area has been set aside in formally restricted zones..." [211]

The plan at the U.N. is well advanced. They are taking over a bit at a time, while we watch reality TV, obsess over the latest Apple product and plan our next vacation. Elizabeth Nickson reports : [212]

"The Global Biodiversity Assessment Report [213] listed the following things as unsustainable: private property, single-family homes, paved roads, ski runs, golf courses, logging, plowing, hunting, dams, fences, paddocks, grazing, fish ponds, fisheries, drain systems, pipelines, pesticides, fertilizer, cemeteries, sewers, and so on."

"In 1993, the EPA circulated a detailed action plan on how, over the next eight years, the U.S. environmental regulations would conform to those of the U.N. "Natural resource and environmental agencies...should...develop a joint strategy to help the United States fulfill its existing international obligations (for example, Convention on Biological Diversity, Agenda 21)..the executive branch should direct federal agencies to evaluate national policies...in light of international policies and obligations and to <u>amend national policies</u> to achieve international objectives.""" [214] (Emphasis added)

Maybe we now have an answer for Barak Obama's refusal to approve the Keystone XL pipeline, and his use of the EPA to destroy the US energy sector, at a time when more energy at lower cost will be essential to avoid the sort of 'fuel poverty' that has been created in the UK by 'green' energy policies.

Maybe Americans have to start dying from cold and cold related diseases before the collective 'light bulb' will go on. Unfortunately, it takes years to rebuild or refurbish energy generation capacity, years that have been wasted waiting for global warming. New non-"green" energy sources such as thorium fueled reactors[215] have been put off for years, and there will be huge time lags to bring that electricity to market. During those years people will die because of these failures.

The battle over global warming theory is almost done, but now all the money has been spent, the world economy teeters on the brink of collapse, China's economy is on the skids, 46,000,000 Americans, in what was the middle class, have been reduced to food stamps, the E.U. has one functioning economy: Germany, the world is awash in non-money, printed at the whim of the U.S. and E.U governments, that can be borrowed for 2% or less, and the U.S. government is moving at a record pace, destroying the U.S. energy industry. We have a new battle to win, a real battle, with real casualties: The battle with hunger, disease and cold. [216]

Fortunately there is a 'silent majority' of people in the science business who are skeptical of the global warming crisis. It is a pity they did not speak up much louder and much sooner. [217]

Conclusion.

The coming New Little Ice Age was foretold in the movement of the planets and records of sunspots. Anyone could have seen it coming, not just high priced researchers with PhDs. It has been a horror story, repeated at regular intervals of starvation, death, malnutrition, disease, collapsed and disrupted civilizations, and is recorded in many cultures, including ours.

Sunspots have been observed and recorded since the Seventeenth Century, and a connection between low counts and cold has been known since about the middle of that century. Many scientists did see it coming and warned us, or attempted to, however there were special interests that got in the way of the warnings.

There were too many people, corporations, governments, the U.N. and green corporations that fueled the hysteria so as to fatten up on taxes, profits, grants from foundations who fell for the story and membership fees, but now the world is left to starve and freeze, and no one will even admit to the crime. Meanwhile, a record number of billionaires have been created, people who got in on this scam and can now escape to warmer places with their plunder.

Jared Diamond says in "Collapse" at the end:

"What are the choices that we must make if we are now to succeed, and not to fail? (For past cultures) two types of choices seem to me to have been crucial in tipping their outcomes towards success or failure: long-term planning, and willingness to reconsider core values. " [218]

It is now 2015, and the most serious natural threat to humanity since the Black Death is upon us, but we have been denied the possibility of long-term planning, and must act urgently. In our struggle to survive will anyone have the ability to re-consider core values, or will we come out the other end in about 80 years ready to make the same mistakes again?

The U.N. moto is: "It's your world." But how much control do we have over it?

It was the simplest thing: studying climate change and the future of humanity but it was hijacked by the U.N.-IPCC, and those who would tax and control us in the name of 'saving the planet'.

Canadian Lt. General Romeo Dallaire had hoped for a renewed U.N., but after this latest fiasco, the U.N. is headed for the junk pile of history. Who will ever trust the U.N. again after millions starve/freeze to death, while they were pushing 'global warming'?

Because of the gross failures of our governments, the U.N. most of the legacy news media and various non-governmental organizations and green corporations, we now need to focus on saving ourselves.

All the best of luck to you and your family.

Lawrence Pierce. Summer 2015

BACK TO THE LAND APPENDIX

Introduction.

Most people will not be able to escape the cities for the country, and will be forced to make do in houses and apartments, where growing food is not as easy as in the country. But it can be done in back yard green houses, front yard gardens, and indoor 'grow walls.' These solutions will not give you all your food, but they sure will help reduce the total cost to you.

Rabbits, one of the biggest meat producers per pound of food can be kept in a garage or back yard hutch. Do not bother with chickens for eggs, they are too risky, but rabbits should work out far better, and you can grow their food almost anywhere. Imagine growing rabbit food in a pop bottle grow wall, and taking one (or two) bottles down per day and giving it to the rabbits.

You can Google any keywords, such as 'rabbits for food' and have the knowledge of the planet on your desk in seconds. Try: "pop bottle grow wall." This will bring up some great ideas for growing food in any apartment or house.

Start now learning about small lot food production, community gardens, 'grow walls,' green houses and other ways to produce your own food. The Greenland Norse failed to adapt, and died out, but you have the time to learn NOW.

The price of freedom.

Life on your own farm, in the country with firewood and chickens was supposed to allow you to live free, but let's be honest, you are paying for it, just in ways you may not understand right away. The same hidden costs are inherent in city living as well, costs that may never have been obvious to you. Before you sell your condo, or house, buy a truck and a chain saw, and tear out to become a "back to the lander," give some thought to the balance sheet of your life in both places.

Two important things must be understood before making the jump: 1. Do you really want to live in the country, and make a living there, or do you just like the idea of it, and 2. What is your exit strategy if it all goes wrong.

Cities enjoyed huge growth in the late 1800s because of the mechanization of farms, and the expansion of manufacturing jobs in centralized locations. Goods and services always follow people; the consumer society was invented in the last century, and now accounts for a large part of the western world's economy. Living a rural life in the 21st Century does not mean living in the poverty associated with rural living 125 years ago, but it does mean there will be big changes.

The rural school your children go to may not be the same caliber as that Montessori back in the suburbs, so a little extra time needs to be spent on studies at home. However, it may be possible to create better schools with more real learning opportunities, but it will take personal involvement.

If you spent the last few years before the big escape to the country living the good life in the city, you probably won't be thinking about buying land, because you do not have enough money. However some people have inherited from their Boomer parents, the generation that did save, or got a chunk of cash somewhere else, and are ready to move.

Those who eat out a lot, should get ready for a surprise. Restaurants in the country, while offering tasty healthy food, are far from the best of what the city has to offer. The same holds true for grocery stores. There simply cannot be the same variety for the small population base in the country that there is in the city. Every time I am back in Vancouver I hit Meinhardt's deli on Granville St. It actually makes the trip back to the big, evil, stinky, city worth the trip, and Vancouver is supposed to be one of the number two or three, depending on who you believe, best cities in the world, and it is in the Warm Zone.

The act of cutting, and stacking wood for that free heat, may lead to short lived, yet painful back problems, or in a worst case scenario, when your steel toed boots are in town for new soles, a chopped foot, which leads to medical bills, down time, and a feeling of having been really stupid.

Not going downtown to work every day means you are home with your spouse and perhaps children. One of the main joys of your city life included spending 'quality time' with both, but now you spend a lot more time with them, and making it 'quality' gets harder to do.

When the fall rains come, sometimes the well gets a bit of mouse shit washed into it, resulting in nausea, diarrhea, cramps and general malaise. City water has been chlorinated to death, is hard in your mouth, and nothing grows in it, but the farm well, with water that tastes like spring time and moves with an effortless smoothness across the tongue, has put you on the couch for a few days. There is no water meter on the well, and no taxes for municipal water delivery, so the water is free, sort of.

After getting that farm started the tractor breaks down the same week as your kid gets sick, the check you were supposed to get from the guy who bought your last side of beef gets 'lost in the mail,' the dog ate something that makes him puke up yellow stuff, and the left rear tire on the truck goes flat, again you may need some help. The trip to town is another expense, and it is necessary to have two roadworthy vehicles to be sure about that trip.

Hidden costs are the same for life in the city. Usually both partners have to work; the day of the stay-at-home-mom are pretty much over, and if there are children in the family, some after school care must be arranged, and inevitably either mom or dad, or both loses a few days every year to the colds and flu that are brought home from daycare, or school. When the kids are not in school, a nanny has to be hired, and unless both partners are making serious money, one partner is just working to pay for the nanny. City life requires the payment of more taxes, fees, transportation costs, and there is more temptation to spend money. Movies, Starbucks, sushi, pizza, clothes, cars and magazines to mention just a few of life's legal temptations, plus the illegal ones.

Flying for pleasure or business, or just shopping in the right places, with the right card makes it possible to collect air miles. These miles can be redeemed for free flights to places, but of course, there is a price for that as well. Some of these companies will cancel the miles if you do not use them, or to get the flight you want it is necessary to fly when no one else wants to, or it takes hours on the phone or computer to get the reservations. That is the price you pay for the "free" flight.

An absolute necessity in the future, either in the city or the country will be growing as much food as possible. To have a productive garden, you need three things: good southern exposure, water and fertile soil. Fertile soil is the real hard part, but can be done without chemical fertilizers if you want to do some extra work. Compost is essential and animal manure high in nitrogen in the compost will ensure lots of vegetables.

The classic organic method or Indore method was invented by the father of modern organic agriculture, Sir Albert Howard. I don't know if he had been knighted before he figured composting out, but certainly deserved it after.

He had gone to India in the 1930s to straighten the Indian farmers out, teach them a bit of modern farming, and make them better farmers. At least he was smart enough to first see what was going on, and analyze it. The Indians were composting everything, he just refined the process, and took it back to the U.K. He realized that he had little to teach them. He was an unusual Englishman and thanks to this collaboration of cultures, everyone had better gardens afterwards.

The Indore method is complicated, as practiced by Sir Albert, but can be boiled down to this. Build layers of three different materials, use sticks, or boards in the pile, which when pulled out, will allow air into the process. The three layers are: manure; green material, like grass, twigs, hay, and finally some dirt with added ground limestone and crushed phosphate rock. Soak each layer with water as it is built, and put in a new layer of boards, (I use 2x4s) with each layer, placing the boards maximum 5 to 6 feet apart. The pile should be 4 to 5 feet high, and as long as you want, but 8 feet minimum.

Once the pile is built, pull out the boards, and it will heat up, decompose, and provide you with the best fertilizer imaginable. It must be turned two or three times, in intervals of 3 weeks during the summer, to keep it going. The pile will really get hot and the hotter the better, because heat will kill weed seeds but if you miscalculate, the seeds will end up in your garden. The pile should be hot enough that the boards discolor and are uncomfortable to the touch, and watch out, after the boards are pulled, more heat is coming.

There are lots of other good composting ideas in: "How to Grow Fruits and Vegetables by the Organic Method" Rodale, 1961. This book is packed with information on organic farming, and is the one reference book every gardener should have, and is still available on Amazon.

The garden providing the best crops has direct sunshine all summer, from dawn to dusk, lots of water, loose well fertilized soil and no weeds. Of course, this garden does not exist, but variations of it do exist. I am not going to write the definitive gardening book, because, as we say in law, "the field has been occupied," but I will give you my take on what is good and what is not from an organic gardening perspective.

Green manure crops are essential, these are crops that take nitrogen and carbon out of the air, and after you mix these crops with the soil it will be looser, and more fertile. Chemical farmers have to buy fertilizer, but the organic farmer gets if for free, but one thing the chemical farmer cannot buy is humus. It can only be grown. To get yours, grow a crop of winter peas, and fall rye every year, by seeding 4 weeks before the first hard frost, then rototilling the green growth into the soil in the spring, 2 or 3 weeks before planting. White clover is really good to add nitrogen before a crop of corn.

The rototiller is an absolutely essential tool for working the earth, and removing weeds. I really like my Honda mid tine machine which has discs at the end of the tines which reduce the chance of grabbing the crop, or drip line while it works. For a country garden, which has lots of room, lay out all your rows 6 or 8 feet, center to center which will allow the tiller to go right down the middle, tearing out the weeds, then the only weeding to be done is between the plants. If you can't afford a tiller, rent it.

Some bigger plants can be weeded with a 'Dutch hoe.' It is a far more efficient weeding tool than most others, but takes some getting used to. Keep your back straight when you use the hoe, bending over just a bit is the natural thing to do, but results in a terrible back ache, and fatigue. Especially in the spring, fatigue will let the weeds get ahead of you. Cut just below the surface, with forward and backward motions, nipping the weeds but not disturbing the soil much, and use it often.

In the fall, after the crop is harvested, do not get upset about weeds. Unless they go to seed at the last minute, they are working for you because any green material produced is good for the soil when tilled in. A 'weed' is defined as: 'a plant growing in the wrong place at the wrong time.' Any plant can be a weed or not.

During the summer weeding, pack the pulled weeds around the plants to add more mulch which holds moisture in and shades the dormant weeds and their seeds. The only exception is the weed which is near going to seed. If they are pulled out, often there is enough juice left to mature the seeds, so take these away and compost them. The correct approach to any plant in the garden is to ask: 'can this plant help or hurt me,' keeping in mind that the answer to that question can change in a few days, like when the plant goes to seed.

Do not build raised bed gardens, unless you have little space, lots of water, and a small appetite. They really dry out the soil, and all tilling, weed removal etc. must be done by hand, because you cannot get a tiller into it. This type of gardening does have a place in the great green scheme of things, for example, if a gardener is confined to a wheelchair, or is unable to bend down, the raised bed is perfect.

There is another use for the raised bed garden, and that is in places where there is no soil, or there are contamination concerns with the soil that is present. Some folks think there is too much lead in city soil to make a good garden, but a comprehensive soil test will tell you what is there. You will be surprised. There is arsenic, lead, aluminum, zinc, copper and many more scary sounding heavy metals in soil miles from a freeway or any industrial activity. Your own body has the same stuff in it, but the amount is what matters.

Normally they are such low concentrations that they cannot harm anyone, and in fact, they are necessary for life. We are the soil, we came from the soil, and our bodies reflect where we came from, and what we eat. The function of the heart is absolutely dependent on calcium, magnesium, sodium, chloride and potassium, they are electrolytes which regulate the beating of the heart, and other electrical functions of the body, making life possible. Other muscles in the body are able to work because electrical stimulation can be transmitted via electrolytes.

And here is a surprise. After the Russian nuclear power plant meltdown at Chernobyl, the world scientific community got really interested in the effects of radiation on people. Lots of effort was made to follow people who had been exposed to small and large doses of radiation to see how that radiation affected cancer rates and health in general. One of the findings was that a statistically significant group, exposed to low levels of radiation had fewer cancers than the general population. This is not an endorsement for meltdowns, but it does bring into question some of the hysteria about radiation.

Never use black plastic sheet, this is plastic surgery for the garden; it looks good for a few years, then the weeds take over, and just like a returning wrinkle, fixing it afterwards is way harder. I am sure you have seen those 'celebrity plastic surgery gone wrong' faces on the internet. Imagine your garden as the "Cat Woman," it makes weeds look good.

Weeds will grow right through the plastic, and tilling becomes impossible, but if you do till, the plastic will come out in six inch squares, and that takes a lot of bending over to clean up. Rely on organic material, like fresh lawn clippings to shade the weeds, and tilling between the rows to tear out the weeds. Put any kind of mulch: lawn clippings, spoiled hay, fresh weeds, on about 2 feet deep, because when organic material is tilled in, there is no need to pick it back out again.

Never use rocks, plastic edging, or logs for 'borders' for the garden. They just give the weeds two edges, instead of one to grow beside, and tilling is lots harder; the weeds hide the rocks, and you may hit them with a mower, your car, your foot, or the tiller, and damage something.

Definitely get drip line, and focus the water on the crop, not the weeds. Pull the drip line out in the fall, and leave it near the garden. It does not have to be totally drained, small amounts of water inside will not hurt it. Part of the reason I suggest 6 to 8 foot centers on the row is because when running a tiller down the middle, the odds of grabbing the drip line with the tiller are greatly reduced.

Use crushed rock fertilizers like rock phosphate and green sand, and use lime to change the pH of the soil upwards. A soil test will reveal what the pH is, and how much lime you might need to adjust it upwards. For just about any vegetable, a pH of 6.5 is just fine but blueberries like it really acid at 5.5 and corn likes sweet soil at around 7.5. Potatoes should be more acid, say 6.0, this helps keep scab to a minimum. About the only time it is necessary to really lower the pH, with sulfur, is for blueberries, which need an acid soil.

In conjunction with crop rotation, the natural pH of the soil should be considered. For example, if the natural soil pH is 6.0, that would be good for potatoes. Grow them first, but since they are a heavy feeder, plant a green manure crop such as white clover next year, and lime it up to 7.0, which in the third year will be good for corn. Lime takes a few years to run out, so do a couple years of other vegetables in between, before another year of green manure without lime, and that should bring you back to 6.0 or so, and time for more potatoes. Apply compost more heavily after crops like corn and potatoes.

Crop varieties must be rotated to avoid a buildup of pests and diseases in the soil. Always have at least two, and preferably three years between plant varieties, for example, potatoes in year one, and not repeated until year four. Corn in year three, and not repeated until year six.

For those with only a small garden, just rotate from side to side. One exception is peas. They can be grown in the same spot over and over, because they fix nitrogen, and to my knowledge, have no diseases. I much prefer the sugar snap variety because the whole pod is edible, thus no shelling.

Make a map every year of where each item was planted. The garden may look familiar now, but after a winter, things may look a bit different.

Rodale's book has lots of detailed charts, and specifics for pH for each kind of vegetable, but if your soil is loose, has lots of organic matter, lots of nitrogen, and is well weeded and watered, you will get a good crop, even if the pH is not exactly right.

The old saying: "One year's seeding is seven years weeding," is true and it also works the other way. After you keep weeds from going to seed for seven years, you have a (more or less) weed free garden but it is hard work to keep things like dandelions and pig weed from going to seed, just keep trying, it is worth it if you succeed.

Weeds have my undying admiration they are one of our planet's real survivors, right along with ants. The ways they propagate include: seeds that are sprayed out in the thousands; seeds that are attached to a fluffy little sail; seeds that can survive the digestive tract of birds; seeds with spikes that attach to animal fur or pants legs, and seeds that can live in the soil for centuries. Recently I read about Pharaoh beer made from sprouted wheat seeds found in a pyramid. We do not consider wheat a 'weed,' but this story demonstrates the resilience of seeds.

Weeds have other ways of taking over, and they include tops that break off from the root, when pulled, leaving a root that can sprout a new top; roots so long they must be dug out; runners above and below the surface that put down new roots; spikes so sharp they discourage any animal but a goat, and when you try to pull them, require the protection of the heaviest leather gloves.

Weeds also have an almost otherworldly ability to keep growing even after they are pulled out of the ground. All dirt must be shaken from the roots, and if the weed is near having mature seeds, it must be removed from the garden and composted. Finally, they have learned to grow shorter and shorter in response to being mowed. The ones putting out seeds at the end of the season often hug the ground, just low enough to not get mowed again. Dandelions are a perfect example of this last trick.

It is truly amazing that humanity has made it this far.

One strategy I tried to reduce (you can never eliminate) the weeds with the little fluffy sail was a garden vacuum. I bought a gas powered weed blower, and reversed the nozzle so it would suck the seeds up into the bag provided. It really did get a lot of the seeds, and is a useful addition to the organic gardeners arsenal.

I had a dream, one night, after a summer of battling the weeds that a new type of weed had invaded my garden. It was spikey and ugly, and seemed to be everywhere. I tried to contain them by placing clear plastic covers over them, (don't ask me why, this was a dream) but they managed to sprout all over my garden anyway. No seeds, no roots, no runners, nothing. Finally I discovered that they were Wi-Fi weeds. Now, that is a real nightmare.

After the soil has been improved for a year or two, the next decision is what to plant, and when. North America is divided into Hardiness Zones by Agriculture Canada, and the USDA. Start out by reading the seed packages, and starting with a few plants. What you select depends a lot on your climate, and it will change every few years from now on. Keep records of everything.

My first winter on the Moberly Bench, near Golden B.C. some turnips got left in the ground over winter by a previous gardener. We were very strapped for cash, and did not buy a lot of food, so naturally, when the greens came out first thing after the snow melted, they tasted like butter lettuce. That is the only part of the turnip that humans should eat, in my biased opinion. The same can be done with beets; both plants will give you a really early salad, relying on the stored energy in the root. Let the pigs dig up the root.

Start the garden with peas, potatoes, beets, beans and carrots. Then a little later, when frost is not expected, plant spinach, and put out these bedding plants: broccoli, cabbage, leeks, and onions. Finally, after the threat of frost is gone, put out the following bedding plants: tomatoes, zucchini, peppers, and put corn seeds in. Every year try something new, but do not be surprised if something does not work. Usually I have one thing (last year it was carrots) that despite seeding, re seeding and tender loving care, just will not grow.

Mulch everything except peppers, none of the varieties like mulch. Corn is usually grown in long rows, with the prevailing winds blowing across the rows to facilitate fertilization, and if mulch is in short supply, prepare the field by tilling in a big crop of rye and peas, and/or clover which will perform some of the functions of mulch.

Now if you failed to heed my advice about buying land in 'green tomato country' the subject of frost damage must be considered. I have used plastic sheeting at night with success, so on clear spring or fall nights, when you can see the stars, get some 6 mil plastic over anything that might freeze. Do not let the plastic come in contact with the leaves, or the frost will go right through.

Bu my second summer on the Moberly Bench, 1972, the garden was doing well until that fateful morning of August 14. It had been a clear night and I know that it is the sort of night that a gardener might expect frost in late September. I got a killing frost that night and almost everything in the garden was done for. Plastic might have helped, but there was no warning…who would have thought of a killing frost in the middle of August?

Pigs and cows

Caution: my daughter Nikki the vegetarian says the following contains disturbing and graphic content, and reader discretion is advised.

Keeping pigs is a very satisfying thing to do. Buy a couple in the spring, they are called 'weaner' pigs because they have just been weaned from the sow, and will eat solid food, and some milk products. The classic is to have a cow that has just calved, and so has a lot of milk. After the calf is weaned, the milk is separated into cream, and skim milk. The cream is sold, and the skim milk can be feed to the pigs. If you intend to have this much livestock, you are indeed brave, a cow needs attention two times a day, rain or shine, hot or cold, 365 days a year. Besides since a large whack of your tax dollars go to prop up dairy farmers, buy milk and cheese, you have already paid for about half of the cost.

However raising a pig is a summer job, and although we prop up hog farmers to a certain extent, pigs can be fed almost for free. Start the little guys off on commercial pig food, and water and add greens as the summer progresses. Fence a small part of your yard, or garden area that is fallow, and let them root it up. Plant greens or vegetables they can harvest themselves and give them all the non-meat kitchen scraps, all the stuff the dog won't eat.

The 'finish' on a meat animal is the food fed at the end. Often cattle are 'grain fed,' or 'corn fed,' and the whole purpose is to put on weight, and give the meat a certain flavor. Corn is another way to finish a pig. We have an epidemic of fat people because of corn, and corn sugar, and the pig will get fat too.

One year, I finished two pigs with apples. This makes very lean and tasty meat, and since most places in the country have old apple trees that still bear, the fruit is there for the taking, and right when needed: in the fall. Apples need TLC during the year or they will develop scale, scabs, or other blemishes which makes them undesirable for us.

A pig will eat a half rotten apple, or one with worms, or vegetables out of the garden, and there are even stories of old sows who ate their farmers when they fell down in the pen. But not to worry, a yearling pig won't eat you, the biggest risk is that they will run away and live in the woods, and then they are hard to catch, and may be bacon for something wild that lives in the woods.

I caught two runaway pigs one fine summer day, with a simple box trap, the kind normally used for rabbits, but made out of a large mahogany tea chest, which was about 3 feet by 3 feet, by 3 feet. These tea chests were surplus at a local tea bag factory and they got used for all sorts of things. With a good coat of varnish they became end tables and book shelves, and out in the shop these boxes were sturdy enough to hold metal parts, junk, pipe fittings, wire and odds and ends.

The box trap was baited with pig food, and placed at the last place I had seen them, which was a small clearing in new woods. Success came after only a few hours, and back to a reinforced pen they went. When small, less than 25 pounds, they can run really fast, but as they get older they slow down a lot. Just like people.

The worm that comes out of an apple will not affect the meat of a pig, but worms in mice, or other critters that might venture into the pig pen and get eaten are a problem. The trichinosis worm is found in many wild animals so it is best to cook all pork you raise at home until it is well done. Trichinosis lives in the muscles of animals, and an older infected animal will have noticeable balls of worm in the muscle tissue. Being safe, by cooking the meat completely is important. Beef is different, because they do not eat meat, unless tricked. When some enterprising farmers gave beef cattle feed containing meat products, the cows rebelled by getting Mad Cow Disease.

Slaughtering animals should be done properly, with a minimum of drama. The first step with a pig is to get an old cast iron bath tub, or similar container ready with lots of boiling water to pour into it.

On the day of slaughter it is essential to not alarm the pig, or stress it out. Care should be taken when moving it, especially for pigs kept in one place most of the time. For the ones who have been moved around all summer, it should be easy. Here is the preferred method. Use a piece of bailer twine, and tie one end to one hind leg, and get a helper to hold onto the other end. The hind leg is about the only part of the pig that something can be tied to; they do not have much of a neck.

Now your job is to lead, or drive the pig to the killing area. Use a bucket of feed, or a push and a pat to get it moving, then make sure your helper does not let go, or you may spend the day chasing it.

Once in the designated spot, it should be approached just like any other day, with a bucket of food, a pat on the head, or whatever is normal. Keep your emotions under control. They can sense them. Not getting attached to or naming livestock is also a wise fence to build and children should be encouraged to treat livestock with respect, but not fall in love with them. This might be hard, little pigs are cute.

Stunning the pig, or steer must be done correctly. In commercial slaughter houses, a mechanical device hits them in the head, or an electric shock is used. On the farm, a gun is the easiest method, but do not even consider a high powered gun, one that might be used when hunting deer. The objective is to stun the animal, not blow it apart.

Use a .22 caliber gun, with a solid bullet, either a rifle or handgun to stun the pig. Make an imaginary X between the ears, and opposite eyes, and shoot for the center of the X. When the pig drops, approach from behind, knife in hand. Grab the pig by the snout, and pull back and stick the knife into the side of the throat, find the jugular vein and cut it.

The jugular is harder to find on a pig, than a steer. It is protected a bit, but if you cut like you intend to cut the head off, which will soon be done, you can find it. Stunning the pig did not stop its heart, so there is lots of blood pressure. Hitting the jugular will be obvious by the rush of blood, no precision is required, and the pig is feeling no pain at this point, even if it kicks. The pig can be left on the ground to bleed, but be careful, they still have some kick left for a while. This is even more so of a steer, which can injure or kill you even <u>after</u> most of the blood is out.

With either animal, test how dead it is by grabbing a rear leg while standing where you won't get hurt if it kicks. Whack the hoof, right where it meets the leg, with the dull back edge of a knife and you will soon know if it is really dead.

The blood can be captured in a pan, the dogs will love it, or if you are European, blood sausage will be on the menu.

Now you need that old bathtub full of really hot water that you prepared in advance. If it is a small pig, two people can hoist it into the tub to let it soak, just like you soak your whiskers (if you are male) in the morning before shaving.

No shaving lather needed, just scrape the hair off with the edge of the knife, after soaking for 10 or 15 minutes. Hold the blade with both hands, at about 75 degrees to the skin, angled towards you. Pull the blade carefully to avoid cutting the skin Eventually the pig can come out of the tub, and be strung up by the heels for gutting. Poke out the skin between the Achilles tendon and bony part of the hind leg and put one hook on each side, insert a short board between the legs, raise the pig with a come along, or front end loader, and cut it open.

Cut the skin open by making a V shaped wedge with two fingers, then after cutting a small hole in the skin of the abdomen, push the abdominal wall down, with the two fingers, and with the other hand, slide the back of a straight blade knife along the abdominal wall while cutting the skin. This exposes the abdominal wall for careful cutting in the next step. When cutting through the abdominal wall, use your hand to push the guts back. Do not remove the skin as it is edible, once shaved and cooked. The point of all this careful cutting is to avoid cutting the stomach or intestines and spilling the contents on the meat.

After the abdomen is open part way, cut out the anus, with the large intestine attached. Tie the intestine with baler twine, or string (which is waiting on your shoulder,) to avoid getting pig shit all over your hams, do the same with the tube coming from the bladder then get the bladder out without spilling the contents. Not feeding the pig the night before will help, and if you are quick, the anus can be cut loose and the attached intestines dropped out without the use of twine.

When removing the liver watch for the green bile sack, and don't cut it open, just carefully remove it. Bury it along with the guts, or leave it out for the birds, they will show up in an hour or two. Keep the liver, heart and kidneys, but if you don't want to eat them, your dog sure does. If the heart is headed for your dinner table cut it open to get rid of the last of the blood.

When the internal organs, above and below the diaphragm are out, and the head is off, the carcass can be split down the middle with a meat saw, and butchered into the usual pieces by consulting a meat chart. Smaller pigs do not need to be split, rather the hams and shoulders can be cut off, and the rest cut up into bacon, and roasts. Make ham by soaking in brine, and smoking; bacon is made by the same method. You can turn both the front shoulder and rear ham into smoked meat, or cook and eat them as is.

Building a smoker is about the easiest thing you can do. Any sort of outdoors confined space, such as a tall, skinny wooden building or an old fridge will serve for a smokehouse. Have racks in the top, or a place to hang large items like hams, a vent for the smoke to leave, and a small smoldering fire in the bottom, or outside with a pipe for the smoke to go in . Most hardwood will be good for smoking. I have used alder, birch, maple, and apple. It should be dry, and keeping it smoldering is the hard part. One way to accomplish this is to take a very dry piece of wood, and boil it in water for 10 minutes. This slows down the burning.

In the southern states, they say that every part of the pig can be used except the squeal. That would be efficient indeed, and people do use the intestines for sausage casing, eat the stomach lining which is called 'tripe,' and cut the head off and boil it for 'head cheese.'

Texas bars in the 1960's all had several things in common. There was the juke box, where Country music takes up about 90% of the listings, the walls were lit up with neon beer signs, and at least 40% of the guys had on cowboy boots or hats, or both. One additional feature that makes the bar complete, along with the beer spigots and mugs waiting to be filled with beer was two large jars. These jars hold about two gallons each, have a white screw off lid, and to the uninitiated appear to have contents that have escaped from a laboratory or the set of a science fiction movie. Usually the fluid in the jar is milky, and just clear enough to make out roundish objects in one jar, and short stick like objects in the other. This is part of the reason Texas bars keep the lights low, to make these objects appear mysterious.

These are 'snacks,' much loved by the beer drinking, good 'ole boy crowd: pickled eggs and trotters. The 'trotter' is the pig's hoof and last joint, it has no meat on it; it is just skin and bone.

For those brave enough to eat one or more pickled eggs, be ready for the repeats. These slippery objects will make your exhaust so deadly that no one will want to be around you, not even your drunk friends. Trotters are pickled pigs feet, the last item before the squeal. Enjoying one of these culinary masterpieces, requires severe intoxication. There is almost nothing good about them.

They remind me of the Sour Toe Cocktail served in the Downtown Hotel in Dawson City Yukon, where a pickled-in-salt human toe is served in a triple shot of hard liquor. Anyone who can drink the Cocktail, making sure that the toe touches the lips, without swallowing it, or puking, gets a frame-able certificate (I have one). Not so in Texas. Eat a trotter, be happy, that's all you get.

I made head cheese one year by boiling the whole head, which produced a heavy gelatin, the meat fell from the bones, which were removed, the brain seemed to disappear, I guess with the bullet, I never saw it again. I did wonder about the lead, however, I understand it can cause memory loss. Sure did for the pig.

What was left was an interesting rubbery, meaty mass that did resemble cheese, a bit. It was rather bland, and not very edible, unless you are really poor, or really hungry, or both, or unless you are determined to make the 100 % most of the pig, which you should be. I imagine there is a way to spice the head cheese to make it more tasty, I just don't know how, and this is not a cookbook.

When killing and butchering a steer the process is about the same, with the following changes. Do not try to tie anything to its hind leg as this will not be appreciated, and may result in injury. Make the imaginary X on the forehead between opposite horns and eyes. When the steer is down grab it from behind by putting two fingers of one hand into the nostrils and pulling back before sticking the knife in with the other hand. The jugular is easy to find. Just pretend that you are cutting off the head, stick the knife up through the soft part of the throat, but do not worry, the steer feels nothing, even if it thrashes around. Be brave, you can do it.

Once bleeding is finished, hoist the steer with the tractor's front end loader or with a pole tripod and come-along. The tendons at the 'heels' will hold the entire carcass, but lifting the animal a bit at a time makes for an easier reach for gutting and skinning.

The nose should be about two feet off of the ground when hoisting is finished, so there will be room for the hide after it is skinned off. Cut out the anus, and tie off the large intestine, then gut the same as a pig, but this animal needs to be skinned; hide is hard to digest. The hide can be tanned, and made into all sorts of leather stuff.

Skinning is tedious. Have a sharp curved knife, and a steel for sharpening nearby for re-sharpening. Pull the skin away from the carcass and slice the connecting tissue carefully so that you do not damage the skin. Avoid cutting hair, it will dull the blade. Start at the top, and work down, to let gravity help pull the skin off and let it fall around the head. Once the hide is removed, sprinkle the inside with rock salt, say 4 pounds for an average hide, roll it up, tucking in all the edges, and put it somewhere cool, until you can get it tanned. Tanning can be done with our without the hair on, but the hide has to be soaked in chemicals before the hair can be scraped off.

Cut off the head after the skin has been pulled right over it and leave as much of the neck as possible. It makes great beef stew.

Normally with a steer, one does not make head cheese, give the head to your waiting dogs who will have fun with it until it stinks really bad, then they will bury it for later. This is part of the reason our relationship with dogs has lasted so long. We find great treats for them, that we really don't want to eat.

Clean up the guts, salvage all the edible internal organs, and bury what is not edible, or leave the rest for the birds. Here on the west coast, the bald eagles will eat dead animals, road kill, guts, etc. They need a lot of help because of the periodic disappearance of the salmon.

There are several uninhabited forest lots behind my farm, and often a bald eagle will perch in the highest branches of a tree, screech a high pitched note, which tells me something dead is nearby. My dogs have come back from these lots with deer legs and hooves on several occasions, when the eagles have shown up. Bald eagles have an incredible sense of smell, and can smell blood within one hour of the spilling of it. Eagles and turkey vultures can strip a full grown deer to the bones within four days of finding it.

Normally butchering is done in the fall, for several reasons. First there will be no need to feed the animal over winter, second they are at their prime after a good summer, and finally, there are fewer bugs in the fall. If possible wait until after the first hard frost has cleaned out most of the bugs.

Wood heat

I have owned lots of wood stoves, since 1974, including two Vermont Castings, and three Regency stoves with fire brick and the Regency is a far better stove. Another great stove is the Blaze King. There are lots of other things you need to know about living with wood heat.

"Draft" is the 'sucking power' of the stove pipe. If the draft is poor, or you get smoke in the house, and slow fires, you probably have a draft problem; cure it, by increasing the height of the chimney. Trees near the house will affect the draft, and since trees grow, things might change over time. Try cutting down a few upwind of the house.

When first installing a wood stove, get someone with some experience to eyeball the height of the roof, and surrounding landscape. If turning on the range vent fan, or bathroom fan, sucks in smoke from the stove, you need more draft, or if the stove is hard to get going, or smoke blows down the chimney when it is windy suspect the draft.

Your stove breathes just like you do so if your house is sealed up, it will use up the air in the house, and the fire quality will be poor, and you may have headaches. Worse yet, cold air will be sucked into the house through every crack and cranny, as it did in my shack in Golden. Control where the air comes in, by having a fresh air inlet right at the back of the wood stove.

I have used stove top thermometers for the last 20 years. Every wood stove should have one, so there is no guessing about how hot it is. Creosote forms the quickest between 60 and 150 degrees F. Many modern stoves or fireplace inserts have blowers to circulate the warm air around the room. Make sure you only run it on 'automatic,' so that it comes on after the firebox temperature gets hot enough, or it will potentially be so cold that the creosote will condense much faster.

The interval between cleanings will be shortest if you have a non-insulated metal chimney that goes through one or more angles into a brick chimney, your stove is an antique that puts out blue smoke, and the firewood is green or unseasoned, and the fire smolders.

As a minimum, with the non-insulated metal chimney and brick outer chimney, clean it twice per winter, and keep a close eye on the non-insulated chimney parts, especially elbows. Creosote is corrosive and can eat these parts up in two years. This was the mistake I made.

If you want to learn more about wood heat, watch the three wood heat and one chainsaw videos I have produced on YouTube. [219]

Planting the garden

Getting all that nitrogen and carbon based humus into your garden in the spring is easy, just get out the rototiller and go to work when the rye is about 3 feet high. If it is left much longer, the stems get really woody, and will plug up the tines of the tiller, but before tilling, spread the compost you made the summer before. Leaving the rye a bit too long is no problem, when the tines get plugged, just keep tilling, stop in one spot, let the tiller dig in, and the friction with the soil will rip most of it out.

To create rows for planting, shovel the dirt into hills, creating a valley between them. Usually the valley was only two feet wide at the bottom, but since I did not use a rototiller for weeds in the city gardens which had limited space, two feet was plenty. Seeds were planted right down the crown of the hill while the length of the hill was as long as the garden, with the normal distance between the crowns of the hills about 4 feet.

My formula for the proper depth to plant was 2 to 3 times the diameter of the seed, but with some things, like lettuce, this was not practical, so I just sprinkled the seed onto the crown, and lightly brushed it into the loose dirt.

The dirt is pressed down firmly on the crown of each row so the seed has good contact with the soil, and the chance of erosion was thus reduced. My weed control system started as soon as the seeds were in. For many years, I just bought seeds from the hardware store, or seed catalogues, but in recent years, here on my third farm, with the encouragement of my Hornby Island partner, Margit, I started saving seeds. There is no comparison between store bought seeds, and seeds that are saved, and planted the next year. The germination rate of seeds drops like a stone after two years, and the seeds you buy may be three or four years old, so between zero and 50% germination is a real possibility. With your own seeds, expect 90%.

Mulch is a necessity for any back yard, or farm garden; there are two principal sources of mulch: spoiled hay, fresh lawn clippings and some people use leaves, but I do not like dried leaves, because they are brown, not green. Green material has nitrogen in it, and you can never get enough nitrogen.

Once planted, the garden must have that mulch blanket thrown over it to hold moisture, fertilize and protect the soil from the corrosive and drying power of the sun. Mulch is another free item for the well informed.

Every farm that grows hay, or has horses or cows, has spoiled hay. The gardener needs hay, not straw; straw is the dry stalk of the wheat plant, which has put all of its energy into the kernels of wheat, leaving just an empty carbon tower that held up the seed head. Straw has practically no nitrogen, and really is only good for animal bedding.

Most farms have a lawn around the house, or open grassy areas that produce green organic material. Using a brush cutter to just demolish green material, and leave it to rot is an unfortunate waste. Why drive around using diesel fuel without a payback? Instead, get a mower with a basket for clippings. The lawn is there for the production of mulch, despite what you may have thought. Also, the idea that grass clipping have to be rotted is another urban myth.

A pile of fresh grass clippings will smell just like cow manure if left in a pile to get hot. Actually, it IS cow manure, without the cow, and maximum nitrogen can be salvaged out of grass clippings by getting it onto the soil quickly. The only reason to process grass through a cow is to get milk and meat, nothing is added by the bovine digestion process, in fact there are losses, mainly in nitrogen rich urine that just splashes away unused. If vegetables are on your mind, leave the cow out of it.

Empty the lawnmower basket right into the valleys between the rows every time it gets full ; just fill it right up to the crown, and stomp it down, so that the only thing that gets any light is the soil on the crown of the row,directly above the seed. The mulch will be mostly rotted by the end of the summer, and by spring, totally gone. About mid-summer pull the mulch back from the side of the row beside your tomatoes and take a peek, you will find fine white roots coming right out of the soil to feed on the mulch.

Spoiled hay should left compacted, just like it comes out of the bailer. Break the chunks off in 'biscuits' about 10 inches thick and leave it compressed, then pack the row with it, so the light cannot get to the side of the hill. The major difference between using spoiled hay, and lawn clippings is that the hay will have lots of seeds. Spoiled hay is like tobacco, once started down that road, it is hard to quit. By comparison, lawn clippings have practically no seeds, and will contribute to a weed 'free' garden after seven or eight years. Grass is not addictive.

When putting in bedding plants, do the rows the same way, and with the exception of peppers, mulch right up to the base of the plant. Be sure to put tomato cages into the dirt deeply, before planting, or the weight of the tomatoes will pull the cage over. Mulch only the bottom of the valley between pepper plants.

Leave lots of room between tomato plants, say four feet, and put them where they will get the maximum air circulation to avoid the black rot which will arrive , in wet climates, around the first week of July, and will destroy the whole crop in a day or two then spread to the potatoes. I have heard that wrapping the stalks of the tomato plant with copper wire will stop the black rot, but have never tried it

Potatoes and squash are done a bit differently than the hill and trench method. With any member of the squash family, like zucchinis, pumpkins, spaghetti and acorn, shovel up a big hill of dirt, about 6 feet in diameter at the base, and 2 to 3 feet high then plant three or four plants on the top. Be sure to imagine the vines running all over the place, getting in the way of the tiller, or other fast growing plants.

Potatoes are easy once you get the soil full of organic material. Do not dig a hole for them, just put the small seed potato on top of the tilled surface, and shovel 8 to 12 inches of dirt onto the seed, and when the plant flowers, "hill" it by shoveling up more dirt around the base. In the fall, you will generally find the potatoes in a two foot radius of the base of the plant; digging with a fork will avoid as much damage as possible. The correct method of digging is to place the fork, at about 45 degree angle to the ground, 2 feet from the center of the plant, push it in slowly, and leverage the soil up.

Gardening is a progressive learning experience. One year may see zucchinis crowding the beans, but next year the gardener will leave a bit more room. Take some photos in the late summer to help your memory next spring.

There are bugs that are really determined, and will eat what appears to be a lot, but consider, for a moment, the strategy the plant uses. A cabbage is a good example; it puts out large numbers of leaves before it gets around to making a head. In effect, it is getting ahead of the bugs, by leaving all those leaves for them.

Get good levels of humus in the soil, do not turn it over in the spring, or fall, make compost every year, grow a green manure crop, mulch, and you will have great crops. Hilling is important in a garden where water will pool for more than a day after a hard rain, but in drier places, just plant seeds and bedding plants in flat rows, and water with a drip line. Hill the potatoes and squash, and rejoice if tomatoes do not rot, because in wet climates they must be grown near a heat sink, like the south side a house, and covered when it rains.

The weeds that do sprout near the crown of the row should be pulled early, and added to the mulch. Until they go to seed they are working for you as mulch.

Weeding is reduced to that area right around the crown of the row and the working area is nice and clean, with fresh hay or lawn clippings to kneel on. Plants should grow rapidly, so do not worry about pests. The plant, being well fed, will outgrow any damage done by insects, even the cabbage butterfly larva has a tough time keeping up. Aphids do, on occasion get out of hand, but in a small farm or city garden, they can be squished by hand, as can slugs if you forget to buy slug bait.

Slugs reared their ugly 'heads' again in my life after moving to the wet coastal climate of B.C. They thrive where it is wet, and mild in the winter, and Vancouver, like San Francisco, has it all, plus miles of lawns which are the best buffet a slug can get, unless of course they find the vegetable garden. I have tried every possible trick to defeat these slimy juggernauts, including sand traps, beer traps, salt traps, and hand picking, but the absolute best is "Corry's Slug and Snail Death." Invented by an Englishman in the mid 1800's which is now produced by a company in North Bend Washington. You will note it is called "Death," not "repellant."

This stuff sucks the water out of slugs faster than a mile of Texas pavement in the summertime. The organic gardening gurus eschew this product, because it contains metaldehyde, however I am not sure why. Metaldehyde has 8 carbon atoms, 16 hydrogen atoms and 4 oxygen atoms. It does not contain any chlorine atoms, or phosphorus atoms. Chlorinated hydrocarbon based pesticides have chlorine atoms in them, while Roundup contains phosphorus. This is a good sign that Corry's will not build up in the soil, or wreak havoc on the reproductive habits of weeds. While it is an irritant to our respiratory system, it is a far bigger irritant to the slugs that would demolish your garden. Besides, who would snort it?

Lay down very thin layers in bands around the garden, keep your dog or cat out of the garden for a few days after applying the poison, and look every day for corpses and put them in the garbage. Secondary poisoning of birds, snakes, frogs and hedgehogs is by far the biggest risk of this product, because they all eat slugs.

Restoring Soil

Every inch of farmland in North America has been assessed for soil quality. Normally a map is available that has details of soil types, the extent of good soil, and lots of other good information, and usually it is free, or low cost. Provincial Ministries of the Environment have them for each Canadian province, and the USDA is the source in the U.S.; try a Google search for "soil map" for the area you are interested in. There are also contour maps that give clues to what might be beyond a small forest, or behind a hill, and what the slope of the land is on neighboring properties.

Soil maps are available for cities as well. New York City has one! I guess there must be some soil under all those buildings and pavement.

Even though your dream farm may be in a good soil zone, the soil may be run out. If goats have been on the land, it may be destroyed. There are also anomalies, within any zone, so get the soil tested. The test must include: nitrogen, phosphorus, potash, organic material, and pH. Every one of these can be adjusted, by the addition of organic material, (by growing it) rock phosphate, (adds phosphorus,) green sand (adds potash and trace minerals,) and lime (for raising pH,) or sulfur (for lowering pH).

pH is the measurement of hydrogen ions in the soil. The test reveals how acid or alkaline soil is, pH ranges from a low of 0 to a high of 14. 7. Any number below 7 is acid. Any number above 7 is alkaline. Every plant has a pH value where it does the best.

The English Professor I bought my farm from was not a farmer, and gave the hay on the field away every year, he was essentially strip mining the soil. It would have been far better to cut the hay, and leave it on the field to rot, but he wanted to be a good neighbor, so he let a local retired engineer, with a few cows, take it away. In a normal farm operation, the cow manure would be returned to the hay field every year, and the net loss would be much lower.

There are many other trace minerals that will be reported on a soil test, and if you get the deluxe version, and tell the lab what crops you want to grow, they will advise how much of each mineral to add to the soil. Unfortunately, they give the amounts in the chemical fertilizer jargon, and some conversion to organic is necessary.

Organic farming is easy. It is hard to put on too much crushed rock fertilizer, because it breaks down slowly, all the organic farmer is doing is opening the buffet, and allowing the plants to eat what they want. Chemical farmers give the plant what they think the plant needs, which is sort of like force feeding geese to make foi gras. Chemical farming sometimes leads to crop damage, because too much fast acting fertilizer is applied.

Part of the culture of chemical farming is akin to the 'blood work' your doctor orders. Doctors, just like chemical farmers, are addicted to tests, they cannot look in a person's eyes, or at their tongue, or poke around a bit and figure anything out. They have stopped trying. Patients, like plants are tested for everything. Diagnostic leaf testing is done to determine what is in the plant's system. Many organic farmers can look at a plant, or smell it, and tell what is going on.

A visit to a big city library section on 'farming and gardening' will reveal hundreds of books on chemical farming, and a few on traditional or organic farming methods. There is much lost art to recover; maybe you could work on it.

Starting in the spring of 2004, I began restoring the soil on my Hornby Island farm. This process would continue for three years, until I had increased the fertility and organic material to the point where I could begin planting. This step was far easier before attempting to plant anything, but could have been done between the rows in an established vineyard or other plantation.

Over the six acres where I ultimately planted blueberries and grapes, I started with crushed rock phosphate and green sand which are the two main fertilizers in organic farming. They needed some 'breakdown time,' so that is where we started. Margit was a big help, operating the 1941 Farmall tractor that came with the place. I did not want to tie up a lot of cash in equipment right away, so I built a fertilizer spreader out of odds and ends from the Recycling

Depot. This contraption mounted on the back of an old wagon which was pulled around the field by Margit on the tractor. The spreader worked just like the typical suburban fertilizer spreader. It had a hopper, a wheel that drove a shaft which agitated the fertilizer, and holes in the bottom to let it fall out.

I was in the wagon, with up to 20 bags of fertilizer, which I opened and poured, one by one, into the hopper of the spreader, as the wagon bounced around the field. One wheel on the wagon drove a bicycle wheel on the spreader which was connected to the shaft that agitated the contents of the hopper.

It is nice to hang around with a woman who, like many of our generation, started driving with a clutch and a gearshift; Margit was a natural with the Farmall.

The rate of application was about 1 ton to the acre of rock phosphate, and 1/2 ton to the acre of greensand. Rock phosphate contains about 30% total phosphoric acid, but most is classified as 'insoluble' and differs from commercial 'super phosphate.' Crushed phosphate rock is what super phosphate is made out of, and is much slower to break down in the soil, which means no charts to study, or precise rates of application to worry about, and no chemical plant kill.

Green sand contains about 3 % soluble potash, but lots of other trace minerals. After these amendments were applied, I surface tilled with discs, and did subsoil loosening with a 2 foot ripper.

This is not fast food for the plants, but slow food.

The standard classification of any fertilizer, organic or chemical is done with three numbers, such as 3-5-10. These numbers show the percentage by weight of each item. The first number shows the amount of nitrogen (N) the second, phosphorus (P) and the third, potash (K). A 10 pound bag marked 3-5-10 would have 3 pounds of nitrogen, 5 pounds of phosphorus, and so forth.

After spreading the crushed rock fertilizers, two loads of chicken manure were spread on the surface. Chicken manure is a fantastic source of nitrogen, and a by-product of the poultry industry, which grows 'broilers' or meat chickens in big barns. The chickens are free run, and live in a carefully controlled environment; heat, humidity, special food, and light are all regulated to ensure rapid growth and the floor of the barn is covered in sawdust so in just six weeks after the little chicks have been introduced to this world, they are ushered out to take a trip to the 'processing' plant. One such plant in Qualicum, B.C. is called: "The Cluck Stops Here."

Once the broilers are gone, the sawdust with the high test droppings are scooped with a Bobcat, into waiting trucks, and driven away to farms all over the Province. I bought 200 cubic yards (which was delivered in two 40 foot trailer loads) and spread it on six acres. It comes out of the truck bone dry, so it is easier on a manure spreader than heavy wet cow manure and bedding. A dust mask is a good idea, and possibly goggles because apparently there is some eye disease associated with chicken manure.

The first year, before addition of the fertilizer and chicken manure, I tried to grow some fall rye, but it only reached 18 inches by early summer. After doing the soil improvements, I planted fall rye, and in the spring it was over six feet tall. One stalk hangs on the wall, in the kitchen, and at 77 inches is taller than me, my kids, or anyone else who has ever come to visit. In the spring this crop of green manure was surface tilled in with the disc. I planted fall rye and peas again in 2005 and 2006, and worked them into the soil.

The last crop of rye in the vineyard, hit an average of 6 feet, with no additional fertilizer during the last 5 years.

If there are no chicken operations nearby, look for horse stables. Often people with too much money buy horses, and have trouble getting rid of the manure. This is where you come in. Using fresh horse manure on the garden is great, just be sure to compost it, it is hot! Big fields are a bit different. The sheet composting method can be used with any manure. Farmers spread cow and pig manure in the spring, before the grass starts to grow, and anyone trying to restore soil can do the same, all it needs is a bit of surface cultivation, with the discs or rototiller pulled behind the tractor.

The process of improving my soil brought me into conflict with one of my wild neighbors, the Canada Goose. On an open field, sprinkling fall rye around is sure to attract their presence, more than putting it in the rows in a vineyard. The posts in the vineyard appear to discourage them, because of reduced 'runway' opportunities, however out in the open no such problem for them exists.

On one trip to the farm, in the fall of 2005, I found that 100% of the two bags of rye that I had seeded had been exchanged for goose shit. The problem was what to do about it since I was not living here yet, did not have a dog, and the rules about shooting or trapping them were so backwards that there was not much that could be done.

Hundreds of geese visited the farm that year and the next, doing lots of damage, and while I was able to shoot a few during the hunting season, it made only a small dent in their zeal to eat my grain. On one occasion, I shot one, and left it on the field, hopefully as a warning to the others. No such luck, they came back, in full force, and just strutted and pecked their way around the corpse, keeping a decent distance of about two feet.

I am sure you have heard that emotional story about how the Canada Goose mates for life. Some folks push that as a sign that they are higher beings. One lady here on Hornby actually talks to them, and offered to 'talk' them out of eating my seeds (it did not work).

Let me tell you about how attached these things are to each other. Many times I would shoot one member of a flock, and the rest, save one, would depart with a great flapping and honking. The one who stayed was the mate, and he/she would observe a period of mourning, usually lasting about 3 minutes, then would go right back to eating, within a few feet of the corpse of his/her beloved.

I am sure there are lots of academic papers on how bird flocks are formed, and why, but one observation I made convinced me that some flocks are made up of the losers. One such loser flock of 12 Canada Geese and one duck visited frequently in 2005, to eat grain, and each time they came, I went out and shot one or two members. I did not bother with the duck, but probably should have, because, in retrospect the duck may have been in charge. This flock kept coming back to the same place on my field, over and over, until I had killed seven of the geese. Each time they would just land, and walk around the bodies of their loser friends eating, until I came out to see them. When finally reduced to less than half of their former strength, they left, never to return.

By comparison, some of the flocks that do migrate through will suffer one loss, then move on. Most of the geese here are the non-migrating variety. Their ancestors were released from a game 40 years ago to re-populate Vancouver Island, and they never leave.

I did put up a scarecrow which helped a bit, but the best cure was to be on the farm firing the shotgun (this is called 'putting the scare into scarecrow') every day until the rye actually sprouted. I added a length of black plastic pipe to the scarecrow to increase his resemblance to me, with a gun. It did not help.

Organic farming depends on healthy soil, so keeping something growing all the time is critical. In the small garden setting, the soil should never be left bare especially in the summer, when wind, sun and rain will erode the organic material; about the only exception is those short intervals after tilling between the rows to tear out weeds. Don't worry, they will be back.

An exception to the 'no bare soil' rule is that if a large crop of organic material is tilled in, it will form a blanket, and protect the soil. I use this technique in my vineyard, between the rows. Fall rye, tilled into the soil in the spring, becomes just like mulch, keeping the moisture in the soil, and protecting against erosion, and as soon as it rains, the weeds come back, so it is not bare for long.

Some corn farmers in the hot part of Kansas have experimented with no-till mulching during the growing season. Crops of green manure plants have been sown right along with the corn, which has reduced soil temperatures by 30 degrees F, as reported in the "Kansas Farmer" magazine, January 2012.

'Shrinking City' Gardens

The spectacle of Detroit becoming abandoned has left thousands of acres un- occupied, and un-cared for

When restoring any land that has been driven on, had a house or walkways on it, or is in an urban environment, rip the subsoil. It may have become compacted over time. But check for municipal services or power lines that may be within the two foot zone. Houses built in the cities before 1940 often had their own septic tanks, which might still be buried somewhere under the yard. Just fill them in if you run across one. Shrinking City gardeners would be wise to hire a local backhoe operator to rip up the soil intended for the garden. In an hour, usually at a cost of less than $100, a big backhoe could rip up 5,000 square feet. Most of these guys can give a fairly accurate estimate, so ask for one.

Storing food

During my first attempt at farming/going back to the land, keeping food for the winter was mostly about canning. I did a bit of this, and got fairly good at both the pressure canning and hot water bath methods. It is a necessity if there is no electricity.

Now, I think canning is obsolete, deep freezers are cheap, and use very little power. The Ener-Guide sticker that came with a new 20 cubic foot Sears upright freezer that I bought last year says it will use 526 kilowatt hours (kWh) of electricity per year. In B.C. we pay about 7 cents per kWh, so that works out to $37 per year. I don't think one could buy all the lids, heat all that water, and replace the occasional jar for that, besides if you empty the freezer early, it can be unplugged. A well made North American or European freezer should easily last 25 or 30 years, and if Fisher & Paykel, a New Zealand company, make freezers as good as their fridges, they might be a good choice too.

I use plastic bags for freezing everything but juice and soup. Blanching just the right amount before freezing is the key to a good result. Here is how I do it. First, keep in mind that the cooling process should be as rapid as possible, you want the fruit of your efforts cooled, then frozen as quickly as possible The vegetables of your efforts too.

Corn. Cut the kernels off with a knife. Start from the skinny end. The only type that you should attempt to freeze is a heavy variety that is said to be good for freezing. I grow my own variety here, called Worker corn, and it takes about 8 to 10 minutes of steaming, after being removed from the cob. Cool and freeze. Other varieties will take less time but usually when it starts to smell good, that is the moment to take it out of the water. Worker corn will keep 18 months when frozen, and if grown in good soil, will be as filling as steak.

Asparagus, green beans, peas, broccoli, brussels sprouts. Boil, or steam until the color changes from light to dark green, which is about 3 to 5 minutes. All keep especially well if coated with olive oil just before freezing. When unfreezing, just leave out all day, or microwave for one minute, then steam or boil briefly.

Beet and turnip greens, spinach, swiss chard and other green leafy plants, boil or steam until they wilt a bit, then cool and freeze. Every year I ever grew Swiss Chard, I got all excited, and froze a bunch, because it is so prolific. But during the winter those bags of rumpled green stuff do not look to appetizing.

Pears. Peal, and halve, and don't bother with taking out the center, it's not worth the work. Make a solution of water and sugar, say about 1.5 gallons of water and 3/4 pound of white sugar. Boil the pears for about 2 to 3 minutes, or until the edges just start to turn translucent, but if you leave them to long, they get mushy. Do the next batch quicker. Take them out, cool and freeze. I cool everything on flat metal sheets.

The point of the sugar is not to make them sweeter, but to keep them sweet. I am sure you remember from high school biology the notion of membrane potential and diffusion. If you have sugar inside the cells of a pear, and you want to dilute the sugar in the cell, boil in plain water, but if you want to keep the sugar in, boil in sugar water. You do remember don't you?

Apples, peaches, plums, cherries, same as pears, except any fruit can have the peel left on, if you are a REAL foodie. Peaches can have the 'fuzz' wiped off with a rag. Apples should be cored, when in quarters, and the stones removed from the rest.

With all of these things, boiling too long will make the finished product mushy. It will be obvious. Keep in mind that the cooking process continues after you take it out of the sugar water solution. By the end of the first box of fruit, you will be good at it. The mushy ones taste just as good as the firm ones, and the sugar water can be reused if promptly refrigerated after each use, just add more water now and then, as it boils away. Discard when it gets real yucky.

Blueberries, strawberries, raspberries, and blackberries, just freeze on a flat plastic or metal sheet, and later transfer to big plastic bags.

All fruit juice. Just freeze. Save up those two quart milk containers for juice, or even better, the one and two quart rectangular juice bottles with a lid. Just be sure, that for anything frozen in glass or plastic leave some 'head room,' which is a space at the top for the liquid to expand into. Usually 2 inches per liter or quart.

Fish. Freeze in bags of water to reduce freezer burn, and extend the life to one year, however it will keep six months in the freezer by just using plastic. Eat it up, there are more coming up the rivers.

All meat. Wrap in waxed paper, or plastic bags and freeze.

Zucchinis. Cut into rounds, about 3/4 thick, and quarter the rounds (even better, cut the zuk into quarters then cut into rounds). Stir fry in canola oil until the area just under the skin starts to turn translucent, about 5 minutes, then cool, and freeze. If you want bags of ratatouille, add onions and tomatoes, stir fry all together until the zuks are done, and it will all be done. Then cool and freeze. Spices and herbs are added at the end of the cooking process, which keeps the flavor from being cooked away. I personally like lots of hot peppers. Spicy food in Canada gets people's attention. When I first moved here from California, and tasted 'hot' sauce, I knew I was in a new country.

Building the root cellar

You may also want to consider building a root cellar in the basement of your home, or a free standing building outside, if you live in the country. There are lots of books on this subject, and lots on the net, but these are the basics: a sand or clay floor, well insulated walls and ceiling, and a door that keeps the cold out. Built correctly, you can store root vegetables for a year. If you do not want to break up the floor in your city basement to expose the dirt, just be sure the humidity is high, and your basement well insulated and ventilated.

My farm came with a nice west coast style one story wooden house, with lots of glass facing the ocean. As is common here, there was no cellar, so I decided to build a separate root cellar, because every farm needs one to keep root crops, pickles, sauerkraut, and fruit fresh.

The basics of a root cellar are: a building that will not rot, buried on the sides, at least 8 feet deep, and heavy insulation or 2 feet of dirt on top, with a dirt or sand floor, and ventilation. The best location is a hill side so that there is a level entrance, coming back up a ladder with an armload of potatoes for dinner is not an attractive picture.

Once the walls are up, a wooden roof can be built, or a concrete one poured, but wood is easier. Start with a 2inch x 10inch board laid flat on top of the block wall, This top plate will provide something to nail the rafters to, then build a roof strong enough to hold 3 feet of dirt. Use 3/4 inch plywood on the rafters, and get some scrap synthetic carpet from the dump, lay it on the plywood, and cover with a seamless layer of pond liner, and pile on the dirt. I used rough cut Douglas Fir for rafters, and they hold up nicely.

The local garden store should have pond liner in big rolls, which can be cut to the right size. Leave some overhang on the sides, so the rain that soaks through the dirt will flow over the edge. The point of using carpet is to have a smooth surface on the underside that will not puncture the pond liner when heaping on the dirt.

At some point, the outside of the concrete block wall will need to have damp proofing painted on. A perfect job is not important, in fact this step might be skipped altogether, because some dampness is a desired quality in a root cellar. I did paint on this black smelly stuff; I wanted to limit the dampness because the cellar was in a really wet spot.

One or two 10 inch diameter vents should be installed. These guarantee enough air flow, however, the amount of air flow is ultimately regulated with a vent low on the door. In very cold weather the vent can be closed, and opened again when it warms up.

In really cold places there should be an inner and outer door, both insulated. Here, where most winters never see cold weather lower than minus 10 C for a week or so, a single insulated door is good enough, I built a plywood 'sandwich' with 3 or 4 inches of rigid foam insulation in the middle, added hinges and a latch.

The final step is backfilling. Two feet minimum is needed over the roof. I had to really build up the sides to get my tractor to climb up to the roof level.

Your transportation.

Getting your vehicle started at minus 40 degrees.

Make sure that in the fall, your spark plugs, and filters are all in perfect shape. The cold is unforgiving. When engines are worn just a bit, they will have a lot of trouble starting. Any American car with over 60,000 miles may be hard to start at minus 40.

I have a lot of first-hand experience getting mobile when it is cold. During three years in the Rocky Mountains, and a winter in the Yukon, I learned a thing or two about machinery.

Do not start the engine at cold temps unless you really have to use the vehicle. The amount of wear on everything is not worth it. Most engine wear happens in the first 10 minutes. Use a trickle battery charger to keep the battery charged if there are going to be long periods when the vehicle is not used.

Use the proper weight oil. Engine oil comes in several grades, use 5 or 10 weight for super cold weather, then as it burns up, slowly add heavier oil, 20, 30 even 40 as it warms up.

Park your vehicle with its 'back to the wind.' Even the slightest wind on the radiator will cool the engine significantly. Better still, find a garage or barn or other structure to park in. Even a tent will help. Natural heat rising out of the Earth will add a bit of heat in an enclosed space, and there will be no wind to take it away.

Before attempting to start the engine, both battery and engine should be warmed to at least zero degrees F. Batteries can be taken into the house overnight, and if there is no electricity to power a block heater, the best way to warm up the engine is with a pan of hot coals from the wood stove.

Build a big fire in the wood stove two hours before you want to get going, then when there is a bed of glowing red coals, scoop some out and take them outside in a metal container. I used an old steel gold pan when I was in the Yukon. Those hot coals were the only gold that pan ever saw, despite many hours panning at the side of the Yukon and Klondike Rivers.

Slide the pan of coals under the oil pan on the vehicle. The oil pan is easy to spot, it has a plug in it for draining the oil, and it is more or less in the center of the bottom of the engine.

Use a non-flammable support to get the pan of coals right up to within no closer than eight inches of the oil pan, although there is not much chance of anything catching fire at any temperature below zero F have a fire extinguisher handy anyway.

It will take about an hour to warm the oil up, just enough time to have another cup of coffee, or two.

The warmed up battery should be brought out at the last minute before start up. When charging a lead/acid battery that has removable caps, be careful to avoid any sort of spark which might ignite the hydrogen gas that is produced by the charging process. It is unlikely that an explosion would occur unless the battery was charged in a small enclosed space, and you provided a spark, by removing the charger cables without unplugging the charger first.

If you can find some spray ether, sold under brand names like "Liquid Fire" or in Australia, "Start 'ya Bastard," get some. Aerosol cans will stay pressurized for decades, and ether will help start up any type of combustion engine. But be careful, it is very flammable and should not be used near that pan of hot coals.

Follow the directions, and spray it onto the air cleaner. The stuff really works.

Treat you vehicle tenderly at very cold temperatures, because metal becomes brittle, and things like door handles will snap right off sometimes.

Now that you are mobile, make sure you have a kit in the car with: matches, candles, a blanket or two and a flashlight. The candles are to generate a little heat if you end up in the ditch.

To help get out of the ditch, always have a 'come-along', some wire cable and a heavy chain in the trunk. Some modern nylon cables or straps may be as strong as steel. These can be used to winch you out, using another car or a tree for an anchor. Be sure to have something more substantial to hook onto than the plastic loops under your car…they were put there to tie down the car during shipment, but are totally useless for winching. Or buy a four wheel drive with a winch. I did that, and in my youthful enthusiasm got stuck within one hour of coming home with my new beauty. There is nothing like machinery to make one feel invincible.

Slippery road conditions will come and go in the New Little Ice Age, so put two or three heavy duty plastic bags in the trunk, along with a small shovel. If conditions change, and you need some weight in the trunk to improve traction, stop, fill the bags with snow, and don't forget to remove them before it warms up.

When driving, if the snow is heavy, and you start to loose traction at low speed, keep a steady foot on the gas, and quickly move the wheel from the 10:00 o'clock to the 3:00 o'clock position. This will break up the snow pack ahead of your front wheels. When you get stuck on the road, be sure to shovel the snow from in front of the front tires, along with the snow at the back, with that small shovel.

When you get stuck, use the 'back and forth routine'. Put the car in drive, or first, move forward till it stops. Then put it in reverse and go back until you get stuck, then <u>quickly</u> change direction, do the same….wash rinse and repeat until you are out of the ditch.

Methyl hydrate is a chemical that will bind with water. If you pour one cup of methyl hydrate into a container with one cup of water, the total will be 1.9 or so cups. It is useful in your gas tank to get the water out. All gas stations have a bit of water in their tanks, and it only takes a bit to freeze in the gas line, and leave you stranded. All hardware stores have it. Add a glug to every tank of gas. It is called 'gas line antifreeze' at the gas station and is ten times the cost of buying the same product at a hardware store.

Bugs

Have you ever noticed when coming back to the city after time in the country that there are fewer bugs? That is because cities are not healthy for most bugs with the cockroach and bedbugs being two exceptions. Does that tell you something about cities?

The only place in the country I have ever been where there were no bugs was California. In 2002 I drove from Vancouver to San Diego in August. This trip took me along I-5, through the

Central Valley, a huge agricultural area, all the way to L.A. and then down the coast. When I got to San Diego, there were a few dead gnats on the windshield. No flies, mosquitos, black flies, butterflies, dragonflies, moths, grasshoppers, crickets, flying ants, nothing except a few skinny gnats. I naturally expected something else, because of years of summer vacations in the Midwest, driving through farm country, when my Dad's Chevy needed a windshield cleaning twice a day so we could see where we were going. In California the city has come to the country.

Anywhere other than California, get used to bugs in the country. Here are some common problems, and solutions.

House flies, moths and other flying insects that get into the house can spread disease, and be a general nuisance, but fortunately, there is an easy cure: spiders. Just leave the spiders alone, and they will silently catch the flying bugs, wrap them neatly when they are done, and drop them in little piles under their webs. If the engineers at UBC, Stanford or Cal Tech were asked to design a bug catcher/killer, they could never do better than the spider.

Our first summer here, I noticed 'sawdust' coming from the ceiling of one of the back bedrooms. There also seemed to be some web like material that came out of the cracks in this wood beam ceiling, which dangled the wood shavings/chips/sawdust in festive mobiles.

Over the next few days, I found hundreds of fat black ants walking single file up the side of the house, and into a hole, right at the level of the ceiling. A path had been worn in the stain on the siding by all those little feet, and about the same time, I started to see one or two in the house, running from dirty cup to plate on the counter, skulking around the baseboards, or dashing across the floor in the general direction of the back room.

Getting rid of them did not involve tearing out the ceiling, or calling the fumigator. I just got a bottle of Raid Ant Poison, and left it out for them in bottle caps right at the bottom of the wall they were climbing. Any ant I found in the house would be followed around, and a drop or two of this stuff would be strategically placed in front of his 'nose,' which generally caused him to stop and eat it. These ants are worker ants, and they are looking for food to take back to the nest.

When they take this stuff back it will kill the Queen, and the entire nest. Good job workers! But have no doubt ants have existed in their present form for millions of years. I saw ants preserved in fossilized tree sap (amber) when I visited the Dominican Republic and they were identical to modern ants, the tree sap was about 65 million years old. They have evolved into the perfect organism, and the perfect society, and we will be long gone before they are.

Raid Ant Poison contains borax and sugar, according to the on-line information provided by them, and a friend makes her own ant poison with those ingredients, plus a little water.

Wasps will build a 'paper' nest almost anywhere, or some will build nests in the ground. The biggest problem with wasps is that they eat fruit, your fruit. They do have a function in the great eco-scheme-of-things which is to clean up dead animals and bugs, so don't kill them all, just the ones that are 'in your face.'

Here are three ways to deal with them.

1. Build a trap out of a 2 liter pop bottle, and use a hot pointed instrument to melt holes in the bottle, about 4 inches up from the bottom. Clear bottles work best; hang it where the wasps hang around, between 2 and 4 feet from the ground with about 2 inches of fruit juice in the bottom of the bottle. They fly in, and cannot get out, then drown, and hundreds can be trapped near a big nest.

2. In the early hours of the morning, when they are still cold and sleepy, find the underground nest. Pour in some gasoline, and set it on fire. However, if the nest is near a building, use number 1 above.

3. If you are dealing with an above ground 'paper' nest, early morning is the best time too. Put on gloves, a hat, and a heavy shirt, and grab the nest, and squish it, along with its contents. Maybe cover your face with some fine net, although if it is cool in the morning they will have trouble flying. They cannot see well in the darkness, but then neither can you so get a friend to hold the flashlight.

Those wasp sprays made by the big bug killing companies, only work if you get right up to the nest, and squirt it in. So why use them? The risk to you is the same without the spray. I could go on some big rant about how they have dangerous chemicals, and are not organic AT ALL, but simply avoiding the rather steep cost of something that is not needed should convince most people.

The blood suckers, mosquitoes, no-see-ums and black flies are probably the most irritating bugs in the world. Huge caribou herds in the North are driven from place to place every year to escape these biting insects.

Mosquitoes breed in pools of water, including the water in your rain barrel, check it in the summer for swimming larva, and if any are present, get some cheap cooking oil, and put a few ounces onto the surface of the water. The larva will swim to the surface for air, and get a mouth full of oil, which generally puts an end to them.

Once they have hatched and are airborne, the number one way to get rid of them is to encourage swallows to nest around your place. The spiders that are allowed to live in the corners of your house also like them a lot.

There are several varieties of swallow that nest in and near houses; I have green and blue cliff swallows, nesting every year in a hole in the side of the house. Barn swallows, the blue and gold variety, come back every year to nest in a protected spot on the other side of the house, and in the machine shed.

It did not take much effort to get them to nest, I just watched where they were circling around in the spring; they can hover for several seconds while inspecting real estate. There is also a very distinct 'chatter' between the male and female, when discussing possible nesting spots. This is very similar to the discussions I have had with a couple of my wives concerning which house to buy. Listen and you will recognize it, even though spoken in 'Swallow." I nailed a short stick to the beam, and within a day, they were hard at work, bringing mud from the pond.

After they talk about it, they will build a new nest, or refurbish last year's nest, and lay one or two batches of up to 5 eggs. If you intrude into their space after the chicks hatch, you will know it by the sharp chirp they make, and the low passes over your head.

From the time the chicks hatch to being airborne is as little as 12 days. They are fed constantly by both parents, who spend all day in the air catching anything that flies. Chicks know when food is coming; an incoming adult, with a beak full of bugs will land on the side of the nest, greeted by open mouths, cheeping loudly for a share. It seems that the louder the cheep, the more likely they are to get fed. Some get fed, some do not, but on the next pass, the loudest ones are always the hungriest. Usually the chicks are out of the nest, sitting on a wire or branch in two weeks, ready to start life on their own. Meanwhile, mom and dad are busy laying more eggs.

If the adults are disturbed to much just before or after the eggs hatch, they will abandon the nest. It seems they have the instinct to cut their losses, and try again later when they are not being hassled. Baby swallows stare out at the world with beady black eyes and a funny smirking look on their beaks. They shuffle from foot to foot on the edge of the nest looking uneasy, and a bit unsure of the passing humans, just a few feet below them.

Summer days out in the field usually attracts swallows, especially if I am using the tractor. When the grass is disturbed, clouds of little green bugs with see through wings are disturbed, and fly up in clouds to meet their fate. The swallows cut and weave in and out of the clouds, sometimes brushing within a few feet of me or the tractor, catching mouthfuls of these tiny insects.

The cliff swallows are a bit more reserved, and shy of people. They nest in the side of my house, in a hole that should be repaired, but hasn't been because of the repeat guests, is right above the deck. When coming in to bring food to the chicks, these birds glide in, change direction two or three times in a simple approach, and finally hit the side of the hole, fold their wings, feed their young and are back out again in about three seconds. When gliding in they have the shape of a modern war plane; delta wings, angled backwards, fuselage hanging low, the angle of the wing adjustable for better maneuverability, tail spread wide, landing gear up, pilot on full alert for interference from the farmer having a beer on the deck.

An inbound swallow will start maneuvers about 50 yards from the house. From a South West bound approach at 40 degrees, pivot to South, South West, tuck the wings, drop 4 feet, roll to the North, pull up to 20 degrees, wings full out, spread the tail, feet out, hit the side of the hole, fold the wings and go in.

From a West bound approach, wings against the body, a steep dive at 65 degrees, roll to the North, wings wide, gain one foot, veer West, spread the tail, feet out, hit the side of the hole, fold the wings and go in.

They have as many variations on the approach as there are degrees and points on the compass, angles of approach, feet of elevation and directions to turn. In 9 summers here I have never seen two approaches that were the same.

Water systems.

Every place in the country has a well, unless the property is very close to a town. Wells are much different from city water, and need some attention now and then, because, you will get sick from your well. Count on it.

The idea is to reduce the level of sickness and frequency, and by 'sick' I include those irritations to the urinary tract caused by different levels of minerals in the water. These show up when traveling back and forth, drinking water from the tap in town, then the tap at the farm.

There are also water borne pathogens, such as giardia, and leptospirosis, which are serious, but easily dealt with. Often a bit of fecal coliform will get into the water, in the spring runoff, or because of heavy rains. These come from animal feces, and will give you the runs, nausea and cramps for a while, but are really not all that serious.

The e-coli bacterial infections that have killed and injured people are extremely rare, and usually linked to domestic cattle. Jarred Diamond (author of "Guns, Germs and Steel") writes that a large number of diseases historically have come from cattle, so it is absolutely essential to keep your cow, and the manure pile, at least 300 feet from the well, and preferably down hill.

People adapt to slightly contaminated well water, so after drinking chlorinated water in the city, then coming to the farm for a week, you may have some upset from your well water. One well driller told me that there are always background levels of bacteria in most wells, bacteria from deep in the ground, or from the surface, and that the body adapts to them. So if you stay at the farm for a month, your body will adapt, and there will not be any more trouble, unless the bacteria levels rise suddenly.

Water purification with an ultraviolet light filter is the best protection from bugs in the well, and the only caveat is that it does not work as well if the water is muddy. "Muddy" is referred to as "turbidity," and it can be measured as part of a standard water test. The bugs just use the mud for shade, and keep from getting zapped by the light, this is why a particle filter should be installed above, or upstream of the UV filter. The light does not really kill the bugs, it just scrambles their DNA so no reproduction goes on inside you. If you get sick, get some GSE, which is short for 'grapefruit seed extract', and in my experience will put an end to the nausea and cramps.

The water borne illnesses that have taken millions of lives globally, are things like cholera and typhoid fever. They come from sick people, so keep your outhouse 300 feet from the well, and do not let anyone spit into your well, and you will be OK. There is way more risk kissing a stranger at a holiday resort, than drinking from a properly managed well.

Septic systems.

After water is used in your house, it must be gotten rid of. There are two basic systems for household waste, the gray water system, plus an outhouse, or composting toilet, and a regular septic system. I would not even categorize what I had in Golden, it is beyond being a category. Taking water out in a slop bucket is for the most remote cabin, not a house.

The grey water system should have a septic field, but does not need a septic tank, and may not even be legal where you are, you should ask around. The original idea of septic fields was an underground series of perforated pipes, with a bed of rocks where bacteria live. Bacteria will eat up any harmful natural substances and the processed water will flow into the soil, and be gone.

A septic tank is necessary if there is a flush toilet in the house. The tank has two chambers, one where the solids first accumulate, then after they break down a bit, the liquid flows over to the second tank, and from there, by gravity, to the septic field. The world of septic tanks and fields has been invaded by beady-eyed bureaucrats, who have new ideas that they have decided are good for us, and unfortunately, they also get sold on ideas by an industry always ready to find new products to sell.

In B.C. fresh water stored in an above ground plastic tank must be at least 100 feet from a septic tank buried in the ground, even if the septic tank is downhill. Inflexible rules have changed the laws of gravity and physics and made this the only place in the world where water can not only leap out of the ground, but run uphill, penetrating, in the process, the concrete tank below ground, and a plastic one above ground.

Septic fields were, at one time, and maybe still are where you live, as simple as a series of ditches, 12 inches wide, 18 inches deep, with a perforated pipe, and drain rock. They were usually placed near the house, under what was going to be the yard, so the grass got watered and fertilized. But now, they have to be built above ground, with special plastic hoses nozzles and deflectors, laid in a bed of sand, with pressurized water coming from the septic tank. Pressurized water is sprayed upwards through the nozzles, bounces off the deflectors, and soaks into the sand, then into the dirt below. This whole contraption is covered with 18 inches of dirt, and is much more expensive than the original type of field, plus the heaviest thing allowed on it is a lawnmower.

Before starting to build any septic system, find out what the local rules area, because in some places there are big penalties for doing it any way other than the "approved" way, even if it works, and even if it is a mile from a stream, or your neighbor's place. We are all victims of regulatory overkill, but at least there is lots less of it in the country.

The composting toilet was invented for those places where water will not flow out of the septic field into the dirt below it fast enough. Those locations fail the "perc" test. The perc (for percolation) test just measures the speed with which water will disappear and if it doesn't go fast enough, smells and bacteria may get out of the septic tank, and into the yard, which is not a good situation.

Composting toilets come in electric and non-electric, and are significantly cheaper than a full blown septic field. There is also the do it yourself composting toilet, which is just an outhouse in a room in the house, but instead of all the droppings going into a hole in the ground, they fall into a metal box which can be emptied with a shovel, from outside. Normally, after every deposit, a layer of wood ash, sawdust, or leaves is thrown on top. I doubt these are legal anywhere, and if you are tempted to build one, find out what the local code is.

When building a septic tank and drain field , there are several important things to remember: do not flush anything but water, what came out of your body, and toilet paper. Grease from the kitchen is hard on the tank, so put it in the compost. Every week, or two, flush a package of septic tank stimulator, such as Septonic, which is replacement bacteria. The tank needs this stimulation because of the bacteria killing bleach and soap that are sent down the drain on laundry day, and if someone in your household is taking lots of medication, flush Septonic twice or three times as often.

It is also a good idea to lift the lid on the tank every other year or so, take a peek to see if the toilet paper is building up, and if so, push it down, and needless to say, it's best to hold your nose during this operation, and wear rubber boots. Being religious about what is flushed, and adding Septonic will stretch the frequency of the visits by the 'honey wagon.' The last tank I had was pumped out after seven years could have gone on a lot longer.

Finding the tank is easy. It is near the house, in the place where the grass is the most abundant and green.

The outhouse is almost as much a symbol of country living as the tractor, and they are really easy to build and simple to operate. First, dig a hole 6 feet deep. The sides of the hole should be smaller than the frame of the outhouse, so it does not fall into the hole, then the outhouse building is dragged over the hole, and used for a year or two. When the deposits come to within 2 feet of the top, drag the house over the next hole, then put about 3 feet of earth into the hole after removing the house, this will create a mound, which, with the decomposition of the deposits, will bring it down level. Use some of the fresh earth that comes from the new hole to fill the old one.

I never put sawdust or wood ash into the outhouse, but some do, however I do not think it makes any difference. Some people save their used toilet paper in a bucket, then take it into the house and burn it in the wood stove. I have no idea why they do this, maybe they read about it in some homesteading magazine. There is an obvious health risk in keeping used toilet paper around, especially if you have a dog. Dogs just love the smell of all sorts of things that we find offensive, that's why they can bury a cows head for a month, then dig it up and eat it.

I hope you can persuade your employer to pull up stakes in the Cold Zone, and move West or South to the Warm Zone. Good luck in your new life.

Tags

Little ice age sunspots solar wind back to the land preppers survivalist global warming hoax global cooling climate change hysteria ocean acidification hoax green corporations profiteers new little ice age cosmic rays un ipcc i.p.c.c. romeo dallaire solar minimum wolf oort maunder sporer dalton minimum climate reparations rio earth summit climate conference lima climate conference copenhagen paris renewable energy solar wind ontario green crony capitalists wegman ian plimer mann hockey stick Australia Canada carbon tax medieval warm period Kyoto Canada warming 18 years agw year without a summer u s corn belt off the grid fascist environmentalist Elizabeth nickson Henrik svensmark tim ball Lawrence Solomon Robert w felix david Archibald jared diamond brian Fagan donna laframboise john l casey

ENDNOTES

These notes are posted on line at www.LawrencePierce.ca

---

[1] Photo by Phil Carpenter, Montreal December 27, 2012

[2] http://spaceweather.com/glossary/sunspotnumber.html

[3] http://www.drroyspencer.com/2013/06/still-epic-fail-73-climate-models-vs-measurements-running-5-year-means/

http://dailycaller.com/2015/03/02/antarctic-sea-ice-did-the-exact-opposite-of-what-models-predicted/

http://www.friendsofscience.org/assets/documents/CanadianClimateModel.pdf

[4] http://www.crawfordperspectives.com/ClimateKeplerianPlanetDyna.htm p.7

[5] http://www.forbes.com/sites/jamestaylor/2014/04/30/twenty-years-of-winter-cooling-defy-global-warming-claims/

"Cold Sun"   John L. Casey, Trafford Publishing 2011. p. 202

[6] http://sks.sirs.bdt.orc.scoolaid.net/cgi-bin/hst-article-display?id=SNY5703-0-586&artno=0000099481&type=ART&shfilter=U

http://www.dailymail.co.uk/sciencetech/article-2093264/Forget-global-warming--Cycle-25-need-worry-NASA-scientists-right-Thames-freezing-again.html

http://www.washingtonpost.com/blogs/capital-weather-gang/wp/2014/12/04/fall-snow-cover-in-northern-hemisphere-was-most-extensive-on-record-even-with-temperatures-at-high-mark/

[7] http://climate.rutgers.edu/snowcover/chart_seasonal.php?ui_set=nhland&ui_season=1

[8] http://notrickszone.com/2015/04/02/cooling-europe-temperature-vegetation-data-show-

central-european-springs-starting-later/#sthash.QLGC1Ayv.dpuf

[9] "Heaven and Earth global warming the missing science." Ian Plimer, Taylor Trade Publishing. 2009. P. 221.

[10] "The Little Ice Age how climate made history 1300-1850." Brian Fagan, Basic Books. 2000. p.120

[11] http://www.crawfordperspectives.com/ClimateKeplerianPlanetDyna.htm p.3

[12] http://www.landscheidt.info/?q=node/5

[13] Little Ice Age p. 122

[14] "Not By Fire but by Ice." Robert W. Felix, Sugarhouse Publishing. 2000. p.217
[15] Heaven and Earth p. 221

[16] Not by Fire p. 218

http://www.spiegel.de/wissenschaft/natur/schwaechelnde-sonne-loeste-antike-kaelteperiode-aus-a-831401.html

http://www.dailymail.co.uk/sciencetech/article-2141705/Global-cooling-Lake-sediment-proves-solar-activity-cooled-earth-2-800-years-ago--happen-again.html

[17] Heaven and Earth p. 123

[18] Little Ice Age p. 213

[19] "The Deniers" Lawrence Solomon, Richard Vigilante Books. 2008
[20] The Deniers p. 174

[21] The Deniers P. 171
[22] The Deniers p. 162

[23] The Deniers p. 163

[24] http://www.gao.spb.ru/english/astrometr/index1_eng.html

[25] The Deniers p. 167
[26] http://www.financialpost.com/m/wp/fp-comment/blog.html?b=business.financialpost.com/2015/03/27/lawrence-solomon-global-warming-doomsayers-take-note-earths-19th-little-ice-age-has-begun&pubdate=2015-03-29

[27] http://www.cnn.com/2014/02/16/politics/kerry-climate/

[28] Trafford Publishing. 2011.

[29] Cold Sun p. 15.

[30] Cold Sun p. 17

[31] Cold Sun p. 27-33.

[32] http://www.sciencedirect.com/science/article/pii/S1364682612000417

[33] "Twilight of Abundance why life in the 21st Century will be nasty, brutish and short." David Archibald, Regnery Publishing 2014
[34] Twilight of Abundance p.12

[35] Twilight of Abundance p. 18

[36] (http://www.nature.com/nature/journal/v476/n7361/full/nature10343.html
[37] Twilight of Abundance p. 19

Heaven and Earth p. 101

[38] http://www.newscientist.com/article/mg22029400.800-whats-happening-to-the-weathermaking-jet-streams.html

http://www.nature.com/news/2010/100414/full/news.2010.184.html

[39] "The Chilling Stars A Cosmic View of Climate Change" Henrik Svensmark and Nigel Calder, Icon Books, 2007 p. 15

[40] Little Ice Age p.29
[41] http://www.ann-geophys.net/30/9/2012/angeo-30-9-2012.pdf

[42] http://environmentalresearchweb.org/cws/article/news/43667

[43] Twilight of Abundance. p. 25.

[44] Twilight of Abundance p.27
[45] Twilight of Abundance p 21
These notes are posted on line at www.LawrencePierce.ca

[46] . http://www.dailymail.co.uk/sciencetech/article-2093264/Forget-global-warming--Cycle-25-need-worry-NASA-scientists-right-Thames-freezing-again.html

http://news.nationalgeographic.com/news/2011/06/110614-sun-hibernation-solar-cycle-sunspots-space-science/

[47] Cold Sun p.202

[48] Little Ice Age p.118

[49] Little Ice Age p.82

[50] Little Ice Age p.83

These notes are posted on line at www.LawrencePierce.ca

[51] "Collapse how societies choose to fail or succeed." Jared Diamond, Penguin Books. p. 270

Jared Diamond is a professor of geography at UCLA. Among his many awards are the National Medal of Science. His book: :"Guns, Germs and Steel" won the Pulitzer Prize.
[52]"Collapse" 219

[53] Collapse p.266
[54] Collapse p.520
[55] Little Ice Age p.96

[56]http://www.agr.gc.ca/eng/science-and-innovation/science-publications-and-resources/resources/from-a-single-seed-tracing-the-marquis-wheat-success-story-in-canada-to-its-roots-in-the-ukraine-1of11/from-a-single-seed-tracing-the-marquis-wheat-success-story-in-canada-to-its-roots-in-the-ukraine-4-160of-16011/?id=1181305178350

[57] . http://www.grainscanada.gc.ca/statistics-statistiques/cge-ecg/annual/exports-11-12-eng.pdf
[58] Little Ice Age p. 83 -194
[59] Little Ice Age p. 177

[60] Little Ice Age p. 178

[61] Little Ice Age p. 91
[62] Andrew McKillop http://ktwop.com/2013/07/01/beware-global-cooling/

[63] http://www.macleans.ca/society/feeling-cold-be-glad-its-not-1815/

[64] Little Ice Age p. 54.

[65] Little Ice Age p.105

[66] http://www.longrangeweather.com/

These notes are posted on line at www.LawrencePierce.ca

[67] Heaven and Earth p.219
[68] http://www.volcanodiscovery.com/worldwide-volcano-activity/daily-reports.html

[69] Heaven and Earth p.84
[70] Heaven and Earth p.219
[71] Heaven and Earth p. 221

[72] https://www.youtube.com/watch?v=sRLdv8jAhxs

[73] http://thewatchers.adorraeli.com/2015/04/13/sleeping-giant-mount-baegdu-changbaishan-close-to-eruption-north-korea-china/

[74] http://www.volcanodiscovery.com/changbaishan/news/18157/Changbaishan-volcano-China-North-Korea-signs-of-unrest.html

[75] Not by Fire p. 201

[76] Cold Sun p. 77

[77] www.iceagenow.info

[78] https://www.youtube.com/watch?v=XDshr5ILNCw

[79] http://www.bbc.com/news/blogs-news-from-elsewhere-32130414
http://www.cbc.ca/news/canada/newfoundland-labrador/where-s-the-beef-empty-meat-shelves-in-province-s-grocery-stores-1.3014488
http://www.cbc.ca/news/canada/newfoundland-labrador/empty-shelves-low-produce-a-symptom-of-weather-interruptions-1.3006316

[80] http://www.ihsmaritime360.com/article/17440/ice-still-delaying-great-lakes-shipments
[81] www.iceagenow.info
[82] http://boston.cbslocal.com/2015/03/09/giant-icebergs-wash-ashore-on-cape-cod/

[83] http://iceagenow.info/

https://www.youtube.com/watch?v=V_BbpsPaSvk&list=PL00u99IRraJtn38lgAequUjcZR_uFnkdY

[84] https://www.youtube.com/watch?v=IDclbzNVZGI

These notes are posted on line at www.LawrencePierce.ca

[85] https://www.youtube.com/watch?v=RMYUL5D4zg8

[86] http://iceweb1.cis.ec.gc.ca/Prod20/page3.xhtml

[87] http://www.nasa.gov/content/goddard/antarctic-sea-ice-reaches-new-record-maximum/

[88] www.iceagenow.info

[89] Little Ice Age p. 48

[90] Little Ice Age p.50

[91] http://www.weather.com/news/climate/news/warmest-winter-on-record-2014-2015

[92] http://globalnews.ca/news/1857393/was-this-vancouvers-nicest-february-on-record/

[93] http://www.sott.net/article/160827-New-Ice-Age-Interviewing-Geologist-Jack-Sauers

[94] http://www.fao.org/worldfoodsituation/csdb/en/

[95] http://www.theglobeandmail.com/report-on-business/ottawa-takes-aim-at-grain-backlog-with-tough-new-transport-rules/article17367492/

[96] Twilight of Abundance pp. 24-25.
[97] Twilight of Abundance p. 44.

These notes are posted on line at www.LawrencePierce.ca

[98] http://words.usask.ca/news/2011/12/05/research-improves-cold-hardy-wheat/

[99] http://www.crystalinks.com/gmproducts.html

[100] http://www.syngentacropprotection.com/news_releases/news.aspx?id=170465

[101] Twilight of Abundance pp. 45—56.

[102] https://www.youtube.com/watch?v=s3Dk8KOp_XQ&index=1&list=TLpTuf7kdeRs4

[103] http://www.gov.mb.ca/flooding/history/index.html

[104] http://www.cbc.ca/news/canada/calgary/why-alberta-s-floods-hit-so-hard-and-fast-1.1328991

http://www.ec.gc.ca/meteo-weather/default.asp?lang=En&n=5BA5EAFC-1&offset=2&toc=hide

These notes are posted on line at www.LawrencePierce.ca

[105] http://news.nationalgeographic.com/news/2013/06/pictures/130606-flood-rain-europe-germany-czech-austria-flooding-pics-pictures/

http://www.theatlantic.com/photo/2013/06/more-flooding-in-central-europe/100533/

http://europeangreens.eu/news/heavy-rains-cause-flooding-across-central-europe

[106] http://www.dw.de/deadly-floods-hit-germany-and-central-europe/a-16854599

http://www.bbc.com/news/world-europe-22835154

[107] http://www.theatlantic.com/photo/2014/05/balkans-struck-by-worst-flooding-in-120-years/100739/

[108] https://www.youtube.com/watch?v=seOb0R_AwNg

[109] https://www.youtube.com/watch?v=wQTAiRLPPDs&list=PL00u99IRraJtn38lgAequUjcZR_uFnkdY

[110] Not by fire p.216

[111] http://resilient-cities.iclei.org/

[112] http://www.treehugger.com/urban-design/alabama-becomes-first-state-officially-adopt-anti-agenda-21-legislation.html

http://www.democratsagainstunagenda21.com/iclei-when-they-say-local-they-mean-it.html

http://www.thenewamerican.com/tech/environment/item/6945-what-are-the-uns-agenda-21-and-iclei

http://www.postsustainabilityinstitute.org/

[113] http://www.huffingtonpost.com/2012/10/30/hurricane-sandy-cuomo-bloomberg-climate-change_n_2043982.html

These notes are posted on line at www.LawrencePierce.ca

[114] https://www.facebook.com/survivalistprepper

http://www.topprepperwebsites.com/

[115] http://www.macleans.ca/society/feeling-cold-be-glad-its-not-1815/

[116] http://www.bcstats.gov.bc.ca/StatisticsBySubject/Demography/Mobility.aspx

[117] https://www.youtube.com/channel/UCUpEo4Js1Icf3b6cWSMGNBg

[118] https://www.youtube.com/watch?v=acoExDbRlrE

[119] http://www.wnd.com/2012/07/bizarre-chinas-eerie-ghost-cities-arise/#!

[120] http://www.dailymail.co.uk/news/article-1339536/Ghost-towns-China-Satellite-images-cities-lying-completely-deserted.html

[121] https://www.youtube.com/watch?v=HcmyJC5MN9c
[122] http://www.theglobeandmail.com/globe-debate/chinas-african-land-grab/article1367342/

[123] http://www.dailymail.co.uk/news/article-2168507/Footage-shows-brand-new-Angolan-city-designed-500-000-lying-empty.html

[124] https://www.youtube.com/watch?v=lkbUjJG_B0g

[125] http://newsbusters.org/blogs/julia-seymour/2015/03/05/and-thats-way-it-was-1972-cronkite-warned-new-ice-age

[126] http://news.nationalgeographic.com/news/2007/02/070228-mars-warming.html
[127] http://www.metoffice.gov.uk/media/pdf/e/f/Paper1_Observing_changes_in_the_climate_system.PDF

http://www.dailymail.co.uk/sciencetech/article-2217286/Global-warming-stopped-16-years-ago-reveals-Met-Office-report-quietly-released--chart-prove-it.html

[128] Heaven and Earth p.438
[129] http://www.cfact.org/2015/03/02/green-slander/#sthash.ZkDygJyz.dpuf

[130] http://www.treehugger.com/corporate-responsibility/documents-released-greenpeace-discredit-one-climate-change-deniers-favorite-scientists.html

[131] http://www.cfact.org/2015/03/02/green-slander/

[132] Corruption p.213

These notes are posted on line at www.LawrencePierce.ca

[133] http://www.greenpeace.org/international/en/news/features/greenlandmelting170206/

[134] http://www.livescience.com/49150-greenland-ice-loss-underestimated.html

[135] http://www.greenpeace.org/usa/en/campaigns/global-warming-and-energy/science/impacts/

[136] http://www.planetextinction.com/planet_extinction_greenland.htm

[137] http://www.climatecentral.org/news/new-greenland-ice-melt-fuels-sea-level-rise-concerns-17187

[138] http://www.greenpeace.org/international/en/campaigns/climate-change/impacts/sea_level_rise/

[139] Not by Fire p. 203.
[140] http://creation.com/the-lost-squadron

[141] https://www.youtube.com/watch?v=uYl4rHloOPU

[142] Not by Fire p.204

[143] http://www.nasa.gov/content/goddard/antarctic-sea-ice-reaches-new-record-maximum/

http://nsidc.org/
[144] http://nsidc.org/

[145] http://www.ipcc.ch/organization/organization.shtml

[146] http://www.cba.org/cba/newsletters-sections/pdf/2012-04-international_jull.pdf

http://www.huffingtonpost.ca/2012/06/15/canada-fossil-award-rio-20_n_1600627.html?

[147] http://www.ipcc.ch/pdf/press/tor_10_03_2010.pdf

These notes are posted on line at www.LawrencePierce.ca

[148] "The Delinquent Teenager who was mistaken for the world's top climate expert. Donna LaFramboise, Amazon.com 2011 p.69

[149] Teenager p.69
[150] .Teenager p. 72
[151] Teenager pp. 7 , 43.

[152] Teenager p. 7, 10, 16, 44.
[153] Teenager p. 46

[154] Teenager p. 17

[155] Teenager p. 10

[156] Teenager p. 36-37
[157] http://blog.ucsusa.org/michael-mann-responds-to-misleading-filings-in-climate-change-lawsuit-641

[158] Steve McIntyre and Ross McKitrick
[159] https://climateaudit.files.wordpress.com/2009/12/mcintyre-grl-2005.pdf

These notes are posted on line at www.LawrencePierce.ca

[160] Teenager p.52
[161] Teenager p. 66.
[162] Teenager p.67
[163] Teenager p.150

[164] http://www.cfact.org/2015/02/23/world-economy-transformation-tops-un-climate-agenda/

[165] http://www.hrw.org/news/2014/03/28/rwanda-justice-after-genocide-20-years

[166] "The Deliberate Corruption of Climate Science." Tim Ball, 2014 Stairway Press. p. 18. http://www.foxnews.com/politics/2009/07/21/obamas-science-czar-considered-forced-

abortions-sterilization-population-growth/

[167] Corruption p. 27
[168] Corruption p. 49
[169] Corruption p.49
[170] http://www.telegraph.co.uk/news/earth/environment/climatechange/10835291/Scientists-accused-of-suppressing-research-because-of-climate-sceptic-argument.html

Corruption p. 81

[171] "Shake Hands With The Devil The Failure of Humanity In Rwanda." 2004 Vintage Canada Edition. p. 250

[172] http://www.telegraph.co.uk/news/earth/environment/climatechange/10643280/Tony-Abbott-appoints-climate-change-sceptic-to-review-energy-target.html

[173] http://www.chathamdailynews.ca/2014/06/09/canadian-aussie-pms-douse-talk-of-carbon-tax

[174] http://blogs.telegraph.co.uk/finance/ianmcowie/100013377/fuel-poverty-and-cold-weather-the-deaths-that-shame-britain/

[175] http://wattsupwiththat.com/2013/01/07/the-potential-impact-of-volcanic-overprinting-of-the-eddy-minimum/

http://rt.com/usa/166352-us-total-debt-sixty-trillion/

[176] http://www.nationalreview.com/article/388595/robert-kennedy-jr-aspiring-tyrant-charles-c-w-cooke

[177] http://hotair.com/archives/2014/03/18/college-professor-jail-climate-change-deniers/

http://www.indystar.com/story/opinion/2014/04/01/disagree-climate-change-deniers-throw-em-jail/7162971/

http://ecowatch.com/2015/03/16/al-gore-sxsw-punish-climate-deniers/

These notes are posted on line at www.LawrencePierce.ca

[178] http://www.theguardian.com/australia-news/2015/mar/06/coalition-ban-on-second-term-for-labor-appointed-agency-directors-brutal

[179] http://www.cfact.org/2015/02/19/record-keepers-cooked-global-temperature-books

http://www.telegraph.co.uk/news/earth/environment/globalwarming/11395516/The-fiddling-with-temperature-data-is-the-biggest-science-scandal-ever.html

http://www.telegraph.co.uk/news/earth/environment/10916086/The-scandal-of-fiddled-global-warming-data.html

https://stevengoddard.wordpress.com/2014/06/23/noaanasa-dramatically-altered-us-temperatures-after-the-year-2000/

http://www.drroyspencer.com/2015/03/even-though-warming-has-stopped-it-keeps-getting-worse/

[180] http://www.cfact.org/2014/12/22/what-if-obamas-climate-change-policies-are-based-on-phraud/

http://www.americanthinker.com/blog/2014/12/evidence_discovered_that_ocean_acidification_scare_may_be_as_fraudulent_as_global_warming.html

[181] Heaven and Earth p. 332

These notes are posted on line at www.LawrencePierce.ca

[182] http://whatreallyhappened.com/WRHARTICLES/globalwarming2.html

[183] http://www.forbes.com/sites/jamestaylor/2011/11/23/climategate-2-0-new-e-mails-rock-the-global-warming-debate/

[184] http://www.lavoisier.com.au/articles/greenhouse-science/climate-change/climategate-emails.pdf

[185] http://www.liveleak.com/view?i=a80_1248836449

[186] http://www.uoguelph.ca/~rmckitri/research/WegmanReport.pdf P.4

[187] http://www.forbes.com/sites/jamestaylor/2013/02/13/peer-reviewed-survey-finds-majority-of-scientists-skeptical-of-global-warming-crisis/

[188] http://blog.heartland.org/2010/10/huzzah-wikipedia-bans-climate-alarmist-from-editing-global-warming-entries/

[189] Heaven and Earth p.446

[190] http://insideclimatenews.org/news/18032015/fema-states-no-climate-planning-no-money

[191] Corruption p. 283
[192] http://www.noblackislewindfarm.org.uk/index.asp?pageid=604834

http://instituteforenergyresearch.org/analysis/germanys-renewable-energy-transition-misses-carbon-reduction-goals/

http://www.renewableenergyworld.com/rea/news/article/2011/05/making-the-case-for-spinning-reserve-on-the-grid

These notes are posted on line at www.LawrencePierce.ca

[193] https://www.youtube.com/watch?v=P2qVNK6zFgE

[194] http://www.labour.gov.on.ca/english/hs/pubs/oel_table.php
[195] http://www.cfact.org/2012/12/11/green-jobs-mean-red-ink-for-energy-consumers/

[196] http://dailysignal.com/2012/10/18/president-obamas-taxpayer-backed-green-energy-failures/

[197] http://www.breitbart.com/big-government/2013/12/05/energy-subsidies-are-going-to-junk-investments-and-failing-companies/

[198] http://www.judicialwatch.org/blog/2013/07/fed-audit-exposes-obamas-500-mil-green-

jobs-failure

[199] " Eco-Fascists. How radical conservationists are destroying our natural heritage" Elizabeth Nickson, Broadside Books 2012 pp. 38 -39

[200] http://www.fraserinstitute.org/uploadedFiles/fraser-ca/Content/research-news/research/articles/crony-capitalism-lurks-in-renewable-energy-subsidies-program.pdf

[201] http://www.forbes.com/sites/larrybell/2011/08/23/the-alarming-cost-of-climate-change-hysteria/

[202] http://www.drroyspencer.com/2013/06/still-epic-fail-73-climate-models-vs-measurements-running-5-year-means/

[203] http://www.therecord.com/opinion-story/2601599-the-expensive-failure-of-green-energy/
[204] http://sppiblog.org/news/green-incompetence-isnt-sustainable#more-12081

These notes are posted on line at www.LawrencePierce.ca

[205] http://www.zerohedge.com/article/global-warming-exposed-un-funded-fraud

http://en.wikipedia.org/wiki/Green_Climate_Fund

http://unfccc.int/cooperation_and_support/financial_mechanism/special_climate_change_fund/items/3657.php

[206] http://www.thenewamerican.com/tech/environment/item/6946-un-demands-76-trillion-for-green-technology

http://www.thenewamerican.com/economy/commentary/item/3838-copenhagen-un-calls-for-trillions-in-reparations-for-climate-debt

[207] http://www.thenewamerican.com/tech/environment/item/20160-from-1750-ad-un-calculates-300-year-climate-debt-for-u-s?tmpl=component&print=1

http://wattsupwiththat.com/2015/02/18/the-un-climate-end-game/

[208] Corruption p. 44

[209] Fascists p.*39*
[210] Fascists p. 38

[211] Fascists p. 39

These notes are posted on line at www.LawrencePierce.ca

[212] Fascists p.103
[213] https://humanbeingsfirst.files.wordpress.com/2013/03/cacheof-unsustainables_un_global_bio_div_assess_95_pages.pdf
[214] Fascists p. 103
[215] Twilight p.163
[216] http://www.cnsnews.com/news/article/ali-meyer/food-stamp-beneficiaries-exceed-46000000-40-straight-months

http://www.conservativecrusader.com/articles/obama-s-new-epa-rules-would-destroy-our-energy-sector

These notes are posted on line at www.LawrencePierce.ca

[217] http://www.forbes.com/sites/jamestaylor/2013/02/13/peer-reviewed-survey-finds-majority-of-scientists-skeptical-of-global-warming-crisis/

http://www.petitionproject.org/

[218] "Collapse" p. 522

These notes are posted on line at www.LawrencePierce.ca

[219] https://www.youtube.com/channel/UCUpEo4Js1Icf3b6cWSMGNBg

INDEX

1440 year cycle   14

Abbot, Tony   56

Abdussamatov, Habibullo	16, 57
AGW	46
Agenda 21	38, 64
Antarctica	30, 47, 49
Archibald, David	19
Australia	56
Back to the land	70
Ball, Tim	54, 55, 56
Canada	23, 34, 43, 50, 56
Casey, John L.	17
Carbon 14	8, 14
Carbon dioxide	54, 62
Clean coal	62
Climate-gate	47, 58
"Cold Sun"	17
Cold and snow records 2015	29, 32
Cold Zone	44
Corn Belt	34
Cosmic rays	19
Crony Capitalists	62
Dalton	19

Dallaire, Romeo — 56

Du Bayne, David — 27

"Eco-Fascists" — 65

Eddy Minimum — 57

EPA — 65

Fagan, Brian — 15

Fairbridge, Rhodes — 16

Felix, Robert — 38

FEMA — 61

Floods — 36

Glaciers — 22, 49

Golden, BC — 45

Great Lakes — 8, 31, 33

Green Corporations — 47

Green energy — 61, 65

Greenpeace — 47, 48, 52

Greenland — 8, 23, 47, 49

Harper, Steven — 56

Herschel, William — 1

Hockey Stick graph — 59

Holdren, John — 55

ICLEI	38
IPCC	19, 46, 50, 51, 54, 60
Ireland	24
Jet Stream	45
Jump the Shark	64
Kyoto	50, 51
Kukla, George	16
La Framboise	46, 50
Mann, Michael	52
McIntyre, Steve	53
McKitrick, Ross	53
Maunder	23, 26
Manitoba	37
MENA countries	34
Met Office (U.K.)	47
"National Post"	16, 39
Napoleon	57
Nickson, Elizabeth	64, 65
North Atlantic Oscillation	14
"Not by Fire but by Ice"	38
Norse	22

Ocean acidification	58
Ontario Canada	29, 33, 63
Plimer, Ian	46, 58
P-38	44
Proxies	13
Pulkovo Observatory	16
Rio Earth Summit	64
Sauers, Jack	9, 14, 33
Solar cycles	12
Solar minimum	12
Solar maximum	12
Solomon, Lawrence	16, 46
Spencer, Roy	8, 62
Sporer	26
Sunspot cycles	8
Svensmark, Henrik	19
Tambora, Mt.	25
Thorium reactor	66
"Twilight of Abundance"	19
U.N.	19, 38, 46, 50, 56, 60, 63
Volcano	41

"Wall Street Journal"	39
Warm weather records 2015	33
Warm zone	42
Wegman, Edward	60
Wheat	23, 24
Wikipedia	60
Wolf Minimum	22, 25

ABOUT THE AUTHOR

Educated in Texas and Canada, he did undergraduate work in energy and public policy issues at

Simon Fraser University and obtained a law degree from the University of British Columbia in 1983. He retired from his litigation law practice in downtown Vancouver in 2009, after conducting over 120 trials and appeals. During this time he learned how to present a winning argument with the best evidence available.

After retiring he bought a small farm in the Gulf Islands of B.C., intending to grow grapes, but the weather grew steadily colder, and his research into the cause of those changes led to this book.

He lives on Hornby Island, with his partner Margit Lieder and two dogs, Millie and Hanna.

In October 2011 he published a do-it-yourself book on Amazon.com. It has become a best seller, ranking on the Amazon best seller list and selling over 11,000 copies as of spring 2015.

The Art of Fixing Things, principles of machines and how to repair them. Available only on Amazon.com

Made in the USA
Lexington, KY
22 October 2016